Concepts for Understanding Fruit Trees

To Rose and my Collaborators in Science

Concepts for Understanding Fruit Trees

T.M. DeJong
Distinguished Professor Emeritus
Plant Sciences Department
University of California, Davis

CABI is a trading name of CAB International

CABI
Nosworthy Way
Wallingford
Oxfordshire OX10 8DE
UK

Tel: +44 (0)1491 832111
Fax: +44 (0)1491 833508
E-mail: info@cabi.org
Website: www.cabi.org

CABI
WeWork
One Lincoln St
24th Floor
Boston, MA 02111
USA

Tel: +1 (617)682-9015
E-mail: cabi-nao@cabi.org

A catalogue record for this book is available from the British Library, London, UK.

Library of Congress Cataloging-in-Publication Data

Names: DeJong, T.M. (Theodore M.), author.
Title: Concepts for understanding fruit trees / T.M. DeJong.
Description: Boston, MA: CAB International, [2022] | Includes bibliographical
   references. | Summary: "This book provides a clear set of integrative concepts for
   understanding how fruit trees work. The emphasis is on overarching principles
   rather than detailed descriptions of tree physiology or differences among the
   numerous species of fruit trees" – Provided by publisher.
Identifiers: LCCN 2021034219 (print) | LCCN 2021034220 (ebook) | ISBN
   9781800620865 (paperback) | ISBN 9781800620858 (ebook) | ISBN
   9781800620872 (epub)
Subjects: LCSH: Fruit trees. | Tree crops. | Fruit-culture.
Classification: LCC SB354.47.D45 2022 (print) | LCC SB354.47 (ebook) | DDC
   634–dc23
LC record available at https://lccn.loc.gov/2021034219
LC ebook record available at https://lccn.loc.gov/2021034220

References to Internet websites (URLs) were accurate at the time of writing.

ISBN-13: 9781800620865 (paperback)
          9781800620858 (ePDF)
          9781800620872 (ePub)

DOI: 10.1079/9781800620865.0000

Commissioning Editor: Rebecca Stubbs
Editorial Assistant: Emma McCann
Production Editor: Shankari Wilford

Typeset by SPi, Pondicherry, India
Printed and bound by CPI Group (UK) Ltd, Croydon, CR0 4YY

Contents

Preface

In addition to helping fruit growers, fruit tree enthusiasts, students and fellow scientists better understand how fruit trees work so they can manage and study them more effectively, I hope that, by reading this book, the reader gains a greater appreciation of the physiological, structural, ecological and evolutionary capacity of trees to cope with, and adapt to, short- and long-term changes in their environment. This book is the culmination of 40 years of research on trees, analysis of other scientists' research, and teaching university students and fruit growers about fruit tree physiology and tree fruit production.

Although scientists often emphasize the complexity of biological systems—and they are indeed complex when considering details of their genetics, biochemistry, molecular and cellular biology, physiology and ecology—the thesis of this book is that there are a few unifying concepts that make much of the general behavior and responses of fruit trees to the environment or management fairly easy to understand and predict. Presenting those unifying concepts is the primary goal of this book. Thus, this book is not meant to provide exhaustive details about any individual species of fruit trees or detailed descriptions of the genetics, biochemistry, molecular and cellular biology, physiology or ecology of fruit trees. This type of information is found in numerous pomology textbooks. A recent textbook that I would recommend for those details is *Principles of Modern Fruit Science* (Sansavini, S., Costa, G., Gucci, R., Inglese, P., Ramina, A., *et al.* (eds), 2019, published by the International Society for Horticultural Science, Leuven, Belgium).

Many of the examples presented in this book are focused on stone fruit, specifically peach trees, because that was the primary species I was focused on for much of my career. However, I believe that all the general concepts presented in this book pertain to virtually all temperate deciduous fruit tree species, recognizing that specific patterns of behavior will vary by species and cultivar. Most of the concepts also probably pertain to evergreen fruit tree species with some adjustments of sensitivities to environmental signals that trigger seasonal activities; however, this is beyond the scope of this book.

As a person who believes in a creator, I count it a great privilege to have had the opportunity to spend the majority of an academic career studying the nature of the fruit trees as biological systems that are flexibly adapted to acclimate and adapt as required by nature and/or fruit tree managers. Not only did my career provide the opportunity to study the complexity of amazing biological systems, it also afforded the opportunity to work with numerous creative, enthusiastic, dedicated and industrious people of quality. I am indebted to many colleagues whom I have had the pleasure of working with over the past four decades. There are too many to name them all; however, I would be remiss if I did not mention a few who had particular impacts on my research and the formulation of many of the ideas that are contained in this book. These include Jim Doyle and Kevin Day, two of the most knowledgeable, practical pomologists in the University of California system; Carolyn DeBuse and Sarah Castro, pomology research associates; Steve Weinbaum and R. Scott Johnson, pomology faculty colleagues at the University of California, Davis (UC Davis); Eric Walton, Edelgard Pavel, Yaffa Grossman, Mat Berman, Gaston Esparza, Luis Solari, Darcy Gordon, Katherine Pope, Anna Davidson and Claudia Negron, graduate students in my program; Joan Girona, Isaac Klein, Adolfo Rosati, Jordi Marsal, Boris Basile, Gerardo Lopez and Sergio Tombesi, visiting scholars to my laboratory; and Mitchell Allen, Przemek Prusinkiewicz, Romeo Favreau and David Da Silva, postdoctoral scholars and collaborators involved in recent fruit tree modeling work.

Although I did not set out to write a book that summarized my academic research career, much of this book has turned out to be essentially that. The first four chapters mainly deal with how trees take up and assimilate resources (solar energy, carbohydrates, water and nutrients) as well as the general structure of trees. The information in these chapters is largely available in numerous other pomological or tree physiology textbooks. The subsequent chapters that address the distribution and use of these resources for growth over multiple seasons contain novel information and approaches to understanding tree behavior that are based mainly on research conducted with numerous colleagues during my career; this information is available in primary research articles but not yet available in textbooks on the subject.

Introduction

In its essence, biology is the study of how living organisms cope with the constraints placed on them by the physical and chemical laws of the physical world in which they must survive and thrive. Thus, biology is essentially the discovery of biological solutions to solving problems posed by the physical world and other living organisms with which they coexist. This book focuses on concepts for understanding how a select group of biological organisms (fruit trees) are adapted to cope with the major constraints imposed on them by the physical world. In this context I choose to think of trees as problem solvers. Fruit trees are faced with multiple problems posed by physical limitations of their environment, such as variations in temperature, water, light and nutrient availability, and I try to identify some strategies that plants (especially fruit trees) have for solving those limitations in this book. However, in studying and thinking about plants and the problems they must solve, it is important to keep in mind that biology can't change the laws of physics and chemistry, it can only develop strategies to deal with them.

Although scientists often emphasize the complexity of biological systems—and they are indeed complex when considering details of their genetics, biochemistry, molecular and cellular biology, physiology and ecology—the thesis of this book is that there are a few unifying concepts that make much of the general behavior and responses of fruit trees to the environment or management fairly easy to understand and predict. Presenting those unifying concepts is the primary goal of this book. Thus, this book is not meant to provide exhaustive details about any individual species of fruit trees or detailed descriptions of the genetics, biochemistry, molecular and cellular biology, physiology or ecology of fruit trees. This type of information is found in numerous pomology textbooks. A recent textbook that I would recommend for those details is *Principles of Modern Fruit Science* (Sansavini *et al.*, 2019).

Most people who observe fruit trees and wonder how or why they behave in specific ways—such as growing very upright or more spreading, producing many flowers and small immature fruit only to drop most of the fruit later on, growing more on their sunny side than their shady side, etc.—ascribe such

© T.M. DeJong 2022. *Concepts for Understanding Fruit Trees* (T.M. DeJong)
DOI: 10.1079/9781800620865.0001

behavior to the trees as a whole. That is the wrong approach to understanding tree functioning and behavior. Trees are not in control of what they do. The control of what trees do and how they function lies in the individual organs that make up the tree, not in the tree as a whole. The genetic code of a tree only indirectly determines the habit, structure and behavior of a tree by defining the behavioral and functional limits of the organs, tissues and cells that make up the tree. Unlike animals, which have a mechanism for collective control of the whole organism (a nervous system), trees (and plants in general) can be more appropriately considered as collections of semi-autonomous organs that are dependent on one another for resources (water, energy (carbohydrates) and nutrients) but control their own destiny.

An excellent illustration of this principle is what has been described as a "five-in-one" fruit tree that is marketed to home gardeners. Such trees are attractive to home gardeners because they allow the production of different fruits (such as an early-ripening peach, a late-ripening peach, a nectarine, a plum and an apricot) all on a single tree. Such trees are produced by bud-grafting each of these fruit types on to a single rootstock. When this "five-in-one" tree grows, each branch representing the individual fruit types maintains the characteristics of those fruit types as if they were growing as individual trees. The structures produced by their growth (shoots, flowers, fruit, leaves, bark, etc.) and the timing of their development (bloom, leaf-out, pattern of shoot and fruit growth, time of fruit maturity, etc.) remain consistent with behaviors of the same varieties as individual trees. Therefore, the control of a "tree" cannot reside in the tree as a whole but trees must be viewed as collections of semi-autonomous organs that are collectively dependent on each other for resources to sustain their life. This brief book will explain how this view of fruit trees can lead to an integrated understanding of how fruit trees work.

Pomology (the science of growing fruit) has usually been approached by describing different types of fruit trees, their patterns of behavior and how they can be managed to efficiently produce. Tree physiology has generally been approached by describing the structural aspects and behavior of parts of trees and especially the perennial aspects of tree physiology, in addition to details about photosynthesis; respiration; nutrient uptake; biochemistry; water, nutrient and carbohydrate transport; etc., and then attributing the coordination of these processes and the behavior of the parts of the tree to the activities of plant hormones. However, there is rarely an attempt to address the fact that plant hormones are merely signaling molecules and there is little explanation for what or how the signals are initiated or controlled. Thus, there has been relatively little emphasis on trying to develop a set of unifying concepts for understanding what controls how fruit trees work.

I believe there is a general set of integrative concepts for understanding the overall physiology of temperate deciduous fruit trees. The emphasis here is on overarching principles rather than detailed descriptions of tree physiology or differences among the numerous species of fruit trees. Most of what is presented pertains mainly to temperate deciduous fruit trees but some aspects may also pertain to evergreen fruit trees or even trees that grow naturally in unmanaged situations.

Fruit trees are biological systems that have been engineered by nature and managed by humans to harvest solar energy to grow and produce fruit. Although most of our fruit trees have been genetically bred and selected by humans for specific characteristics that make them horticulturally valuable, it is important to remember that most of their physiological traits have been selected by nature to enhance their survival and reproduction in natural settings. In the process of developing strategies for dealing with the limitations of their physical world to survive and thrive on land, there are many issues to confront or problems to solve. It is instructive to continually analyze fruit trees from the viewpoint of understanding the problems they face for surviving and thriving on land, the adaptations they have for dealing with those problems and how they can be optimally managed for desired results.

In general, green plants are nature's original biologically based solar-powered organisms. After the Industrial Revolution and until relatively recent advancements in modern sustainable energy development, much of the energy used to sustain life on earth has been derived from plant-based solar energy collection. The solar energy cells in green land plants are primarily housed in specialized structures (chloroplasts) within leaf cells and leaves are specially constructed to provide a controlled aqueous environment suitable for the function of those solar cells. Furthermore, the primary purpose of plant architectural structure (trunk, branches, stems and shoots) is to bear and display leaves so that they are exposed to solar energy (the sun's rays). Thus, the productivity of any crop is a function of two things: the efficiency of resource (solar energy, water and nutrients) capture and assimilation; and the efficiency of distribution and use of those assimilated resources. Interestingly, while much research has been focused on understanding and enhancing the efficiency of resource capture and assimilation by plants, most agricultural advances have been based on manipulating resource distribution and use. This probably reflects the fact that it is hard to imagine a scenario in nature or in managed agriculture where anything but optimal resource capture and assimilation would be advantageous and selected for. Whereas there are many situations in managed agriculture where optimal distribution and use of resources in an anthropocentric monoculture may differ from what might be advantageous, and naturally selected for, in a competitive natural environment (e.g. plant stature, specific fruit characteristics, time of harvest, etc.). For this reason this book will only provide a cursory description of factors affecting the efficiency of resource capture and assimilation and the bulk of the focus will be placed on understanding how assimilated resources are distributed in fruit trees and how trees can be managed to optimize distribution of resources for horticultural gain.

Understanding the scheme of assimilate distribution within fruit trees allows for a more complete understanding of how fruit trees function and how horticultural manipulation of fruit trees can influence the growth of various organs, as well as how specific organ growth can be optimized to meet specific objectives.

Energy Capture and Carbon Assimilation

2

Fruit and nut trees are not much different from other plants in that they are primarily adapted to obtain resources from their environment and use those resources to grow and produce propagules that can continue their survival as a species. The primary resources that plants are adapted to capture are (in order of importance) light energy, water and nutrients. Without energy a plant could do nothing. At its core, the primary purpose of the structure (branches) of a plant is to display solar energy collectors (leaves) and the primary purpose of leaves is to house and provide an efficient environment for the functioning of solar energy cells (chloroplasts within the green cells of leaves) (Fig. 2.1). Thus, optimal fruit trees and orchards are arranged and shaped for efficient capture of solar energy (Fig. 2.2). Water availability is essential to a plant and is primarily used to keep leaves in a hydrated state so that their solar energy cells (chloroplasts) can function efficiently. The solar energy that chloroplasts capture in their green pigments (chlorophyll) is stored in carbohydrates that are formed through the process of **photosynthesis**. The raw materials for this biochemical reaction are CO_2 (carbon dioxide) and H_2O (water) and the products are CH_2Os (sugars or carbohydrates) and O_2 (oxygen) (Fig. 2.3). CO_2 is obtained from the atmosphere and H_2O is obtained from water stored in the soil. However, the fact that CO_2 needs to be absorbed from relatively dry air surrounding the plant, and that chloroplasts need to be maintained in a hydrated state, poses a major problem for the plant. Leaves are structures designed to specifically solve this problem. They have specialized pores (stomata) that help control the amount of water that is lost from the fully hydrated leaf while CO_2 is being taken up from the relatively dry air surrounding the plant. The primary use of water in land plants is to replace the water that is lost from leaves while they are taking in CO_2 to supply photosynthetic reactions.

Since water loss by leaves through evaporation from leaf surfaces (transpiration) during the process of photosynthesis is the major use of water by land plants, this is also the main driving force for movement of water from the soil water pore spaces, into and through the roots, up the stem, out to the leaves and into the atmosphere. This pathway of water movement through plants is termed

© T.M. DeJong 2022. *Concepts for Understanding Fruit Trees* (T.M. DeJong)
DOI: 10.1079/9781800620865.0002

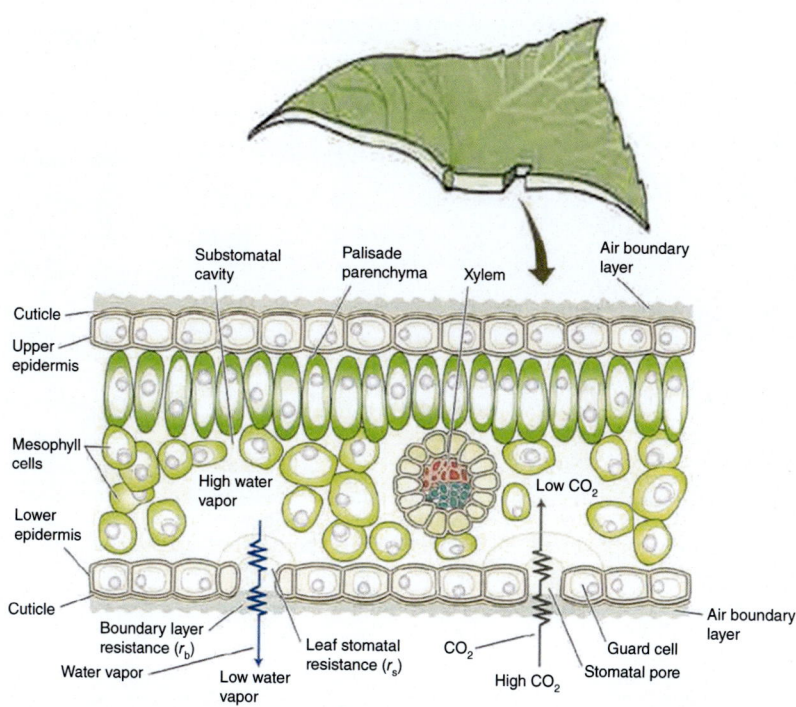

Fig. 2.1. Cross-section of the internal structure of a leaf. The chloroplasts (solar energy cells) are located in the palisade and mesophyll cells. The arrows through the stomatal pores indicate the pathways and resistances for water vapor and CO_2 movement out of and into the stomatal pores of the leaf. (From Taiz *et al.*, 2015.)

the soil–plant–atmosphere continuum (SPAC) (Fig. 2.4). Scientists measure the driving force for movement of water through this continuum in terms of water potential (Ψ). Pure water has a water potential equal to 0.00 MPa and water always moves from a less negative to a more negative water potential. The water potential (Ψ_w) of a plant cell is the sum of two components: Ψ_s and Ψ_p ($\Psi_w = \Psi_s + \Psi_p$). Ψ_s is the solute potential and it is an expression of the tendency of the solute to attract water (like the tendency for salt in a saltshaker to absorb water). Ψ_p is the pressure potential (turgor pressure in plant cells) and it is the pressure that is exerted on the wall surrounding a cell due to the fact the cell contains a lot of solutes that attract water. Ψ_s is always negative because it is an expression of the tendency of solutes to attract water and Ψ_p is always positive in living plant cells and is an expression of the tendency for cells to expand.

When considering movement of water among cells in a plant or between the plant and its surroundings, the difference in water potential between any two points along the continuum indicates the relative tendency and direction of movement of water between those points. For instance, the water potential of open air in an orchard ranges from −10 to −100 MPa, while the water

Fig. 2.2. An aerial photo of a California almond orchard. Trees are planted and pruned to optimally distribute and capture sunlight within their canopies. Thus, these systems represent biological solar arrays. (Photo by E. DeJong.)

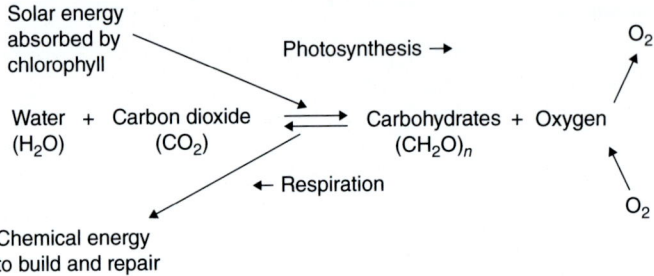

Fig. 2.3. Diagrammatic summary of the fundamental biochemical reactions of photosynthesis and its inverse, respiration. (From DeJong, 1989.)

potential of leaves of adequately irrigated orchard trees can range from −0.2 to −1.5 MPa. Thus, there is almost always a tendency for water to move from the leaf to the atmosphere. Similarly, soil water potential in an irrigated orchard is generally less negative than −0.2 MPa; thus there is almost always a tendency for water to move from the soil through the roots and stem to the leaves. The actual rate of movement through the entire system is dependent on the differences in water potential and resistances to water movement between various points along the SPAC pathway. It is important for growers to manage irrigation so that tree leaves never (or rarely) experience mid-day leaf water potentials more negative than a recommended baseline value during the growing season.

Similar to water use, the majority of many nutrients that are taken up from the soil are used to build and maintain the photosynthetic

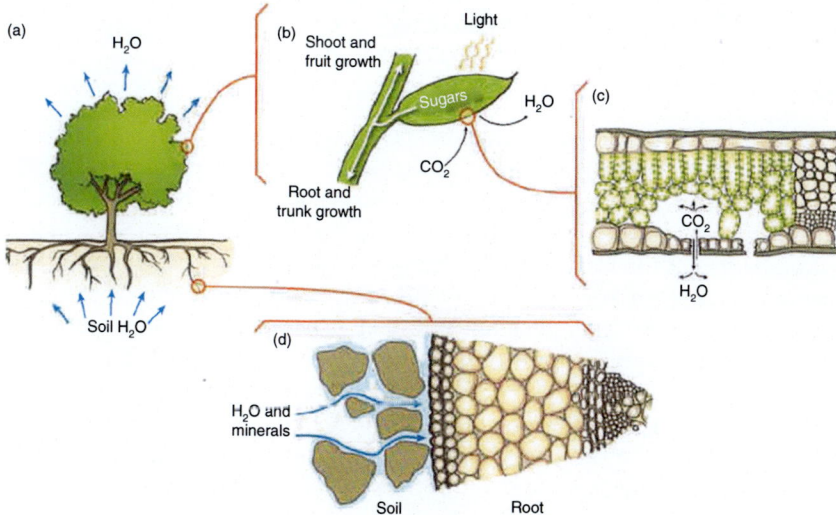

Fig. 2.4. Diagram of the soil–plant–atmosphere continuum (SPAC) of water flow through trees. Water is taken up from the soil by roots (d) and moves through the tree (a) and leaves (b) and out of the leaves through stomata (c) to the atmosphere. (From DeJong *et al.*, 2012.)

apparatus in the leaves and to facilitate energy exchange in other plant parts. Figure 2.3 shows a simple summary of the overall photosynthetic process. The actual process is a complex set of reactions involving many of the nutrients green plants require. For example, nitrogen (N) is a constituent of photosynthetic enzymes and chlorophyll; phosphorus (P) is important in energy transfer processes; magnesium (Mg) is an essential part of the chlorophyll molecule; and potassium (K), iron (Fe), manganese (Mn) and other nutrients play important roles in specific photosynthetic reactions.

Photosynthesis

The photosynthetic process in leaves is dependent on three separate but interconnected processes. A biophysical/chemical process involving the "light reactions" that capture and convert solar energy into chemical energy that drives the "dark reactions" that involve conversion of energy from charged unstable intermediate biochemicals into stable compounds (sugars and starch) for storing and transporting chemical energy around the plant. The latter process involves the use of CO_2 taken into the leaf from the air around the leaf. All three processes—the light reactions, the dark reactions (also called carboxylation reactions) and leaf uptake of CO_2—are required for photosynthesis to occur and represent control points for photosynthesis.

Factors Influencing Photosynthesis

Light

As stated previously, the light reactions of photosynthesis are the processes by which energy from sunlight is captured and converted into chemical energy. Therefore, exposure to sunlight is essential for tree leaves to carry out photosynthesis. Individual leaf photosynthetic capacity varies with the light environment that the leaf is usually exposed to, and leaves on the outside of a canopy have a greater capacity for photosynthesis than interior canopy leaves. Furthermore, an individual tree leaf exposed to full sunlight can usually only efficiently utilize less than one-half the amount of full sunlight for photosynthesis. Therefore, the photosynthetic apparatus of most fruit tree leaves approaches light saturation at less than one-half full sunlight if the leaf is exposed directly to the sun. Figure 2.5 illustrates this point; note that the slope of the line representing photosynthesis begins leveling off at

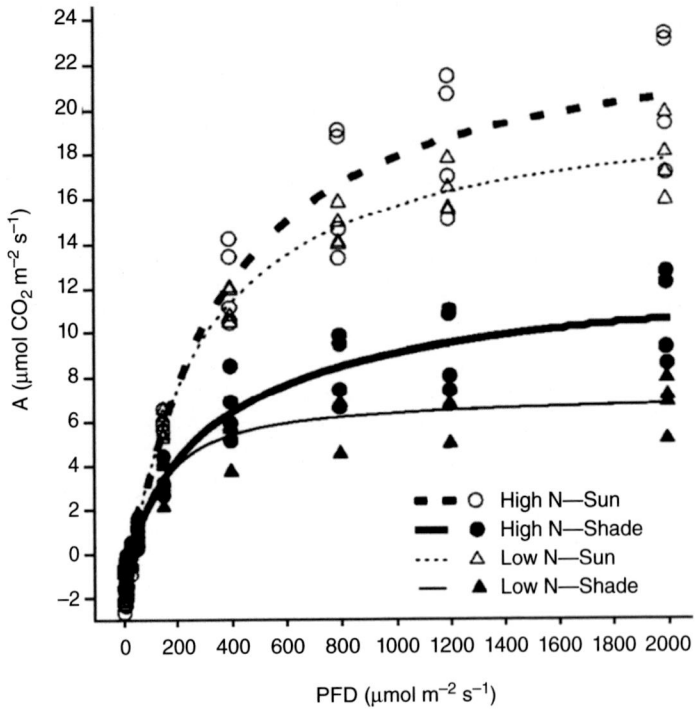

Fig. 2.5. The photosynthetic (CO_2 assimilation, A) response of peach leaves to variable light (photon flux density, PFD) growing in outer (sun exposed) and inner (shade exposed) areas of trees receiving high and low rates of N fertilizer. Full sunlight on a clear day in California has a PFD of about 1800. (From Rosati *et al.*, 1999.)

photosynthetic photon flux densities (PPFDs) less than 800 μmol m^{-2} s^{-1}. Only leaves on the outer surface of the tree canopy are ever exposed to direct sunlight for extended periods of time. Even those leaves are often oriented vertically and/ or are partially folded so the sun's rays strike them at oblique angles. Thus, the leaf surfaces of most deciduous trees are rarely oriented perpendicular to the sun's rays as the sun moves across the sky. Thus, most leaves on a mature tree function for the majority of the day on the steep portion of the light response curve. For this reason, total tree-canopy photosynthesis tends to be directly related to tree-canopy light interception (Fig. 2.6).

Each leaf, located in its zone of the tree canopy, has an ever-changing light environment. Light becomes more limiting to photosynthesis from the outer edge to the center of the foliar canopy. If tree canopies become too dense, the inner parts become so shaded that leaves there cannot carry out an appreciable amount of photosynthesis. Such interior leaves then contribute less energy to the branches or spurs to which they are attached than they need and, eventually, those leaves and the stems or spurs they are attached to die.

Temperature

Photosynthesis is dependent on temperature because the light and dark reactions of photosynthesis involve numerous biochemical enzymatic reactions. Photosynthesis of most temperate deciduous fruit tree leaves functions optimally at leaf temperatures between about 70 and 90°F (20 to 32°C) (Fig. 2.7). Where daytime summer temperatures are rarely less than 70°F (20°C), it is unlikely that low temperatures during the growing season significantly inhibit photosynthesis in fruit tree leaves. However, temperatures above 90°F (32°C) can occur frequently

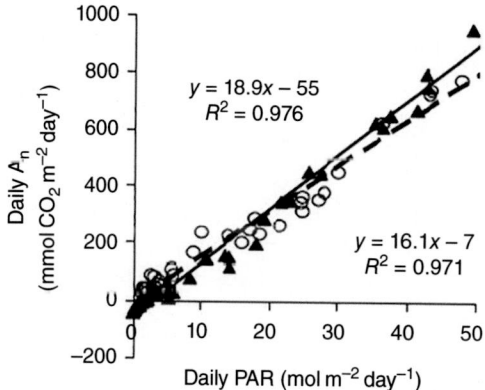

Fig. 2.6. The relationship between accumulated daily light exposure (photosynthetically active radiation, PAR) and the accumulated daily canopy photosynthesis (A$_n$) of non-stressed peach trees grown with high (solid line) or low (broken line) N fertilization. (From Rosati *et al.*, 2002.)

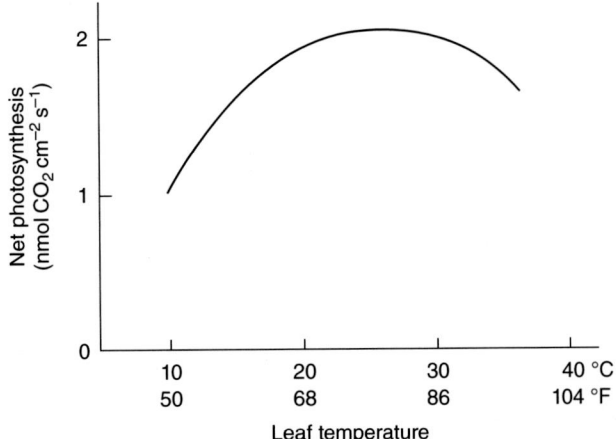

Fig. 2.7. A generalized photosynthetic temperature response of temperate deciduous fruit tree leaves. (From DeJong, 1989.)

in some temperate climates and these temperatures can inhibit photosynthesis, especially when they are accompanied by strong, dry winds.

CO_2 concentration

The concentration of CO_2 in the atmosphere is fairly constant on a seasonal basis but has been gradually increasing over the years due to climate change so that it is now approximately 420 ppm. Ambient CO_2 rarely changes enough on a daily or seasonal basis to affect photosynthesis significantly during the course of an individual season. However, CO_2 must enter the leaf to be utilized in photosynthesis, so any treatment that inhibits diffusion of CO_2 into the leaf by shutting the stomata (such as severe water stress) can negatively affect photosynthesis by decreasing CO_2 concentrations inside leaves.

Nutrient supply

Each molecule of chlorophyll, which is essential for the light reactions, contains four atoms of N and one of Mg. Furthermore, the most important protein/enzyme involved in the dark reactions (RUBP carboxylase) accounts for the majority of N in a leaf. P also plays a vital role in biochemical energy metabolism and transfer. Other elements—Fe, Mn, copper (Cu), boron (B) and zinc (Zn)—regulate enzyme activity. Deficiencies in any one or all of these nutrients often leads to chlorosis (loss of green color) and decreases leaf photosynthesis (as illustrated in Fig. 2.5 for N). Furthermore, a deficiency of any nutrient—for example, Zn—that retards leaf growth also reduces photosynthesis because of decreased canopy size.

Leaf number and exposure

Any tissue containing chlorophyll in the cells, such as young stems and fruit, can carry out photosynthesis. In fruit trees, leaves are the primary organs for photosynthesis. So, optimal exposure of the maximum number of leaves to light normally results in the greatest yield of dry matter as long as fruit set is not limiting.

Water availability

Since CO_2 needs to be taken up from the atmosphere for photosynthesis to occur within leaves and water is lost when stomata open to take in CO_2, trees need to have adequate access to water to keep them hydrated as water is lost during the photosynthetic process. If adequate water is not available, the water potential within the tree will decrease and the tree will begin to become dehydrated. Then, either stomata will begin to close and the rate of CO_2 exchange between the leaves and the atmosphere will be reduced, or the leaves will begin to dehydrate and the light and dark reactions will be negatively affected. In either case the rate of photosynthesis will decline.

Many of these factors interact to result in the pattern of whole-tree photosynthesis that is achieved over a day (Fig. 2.8). As shown for individual leaves, in the morning leaf photosynthesis increases as light intensity increases and generally reaches a maximum level by mid-to-late morning. During the afternoon, temperatures often become relatively high, humidity decreases and vapor pressure deficit becomes greater, plant water potential decreases, stomatal apertures narrow and stomatal conductance decreases, and the photosynthetic rate falls before afternoon sunlight becomes limiting (Fig. 2.8). Note that photosynthetic and stomatal responses are accentuated if the tree experiences water stress (more negative stem water potentials). During warm afternoon hours, photosynthesis is probably limited by temperature or water stress because the stomata begin to close, even though light intensity is still adequate for maximum rates of photosynthesis. However, estimated whole-canopy photosynthesis is usually less affected by water stress during the morning hours (Fig. 2.9) than is individual leaf photosynthesis (Fig. 2.8) because water-stressed trees tend to have higher light interception in morning hours than during the middle of the day because of water stress effects on leaf angles and water stress increases over a day. This indicates the importance of canopy light distribution patterns in determining the photosynthetic behavior of tree canopies. Seasonally, leaf photosynthetic light-use efficiency tends to decrease with leaf age and increased canopy shading both within and between trees even when nutrient conditions are adequate (Fig. 2.10).

It seems logical that, because photosynthesis is dependent on light energy, a tree would have relatively fewer assimilates to carry out growth and

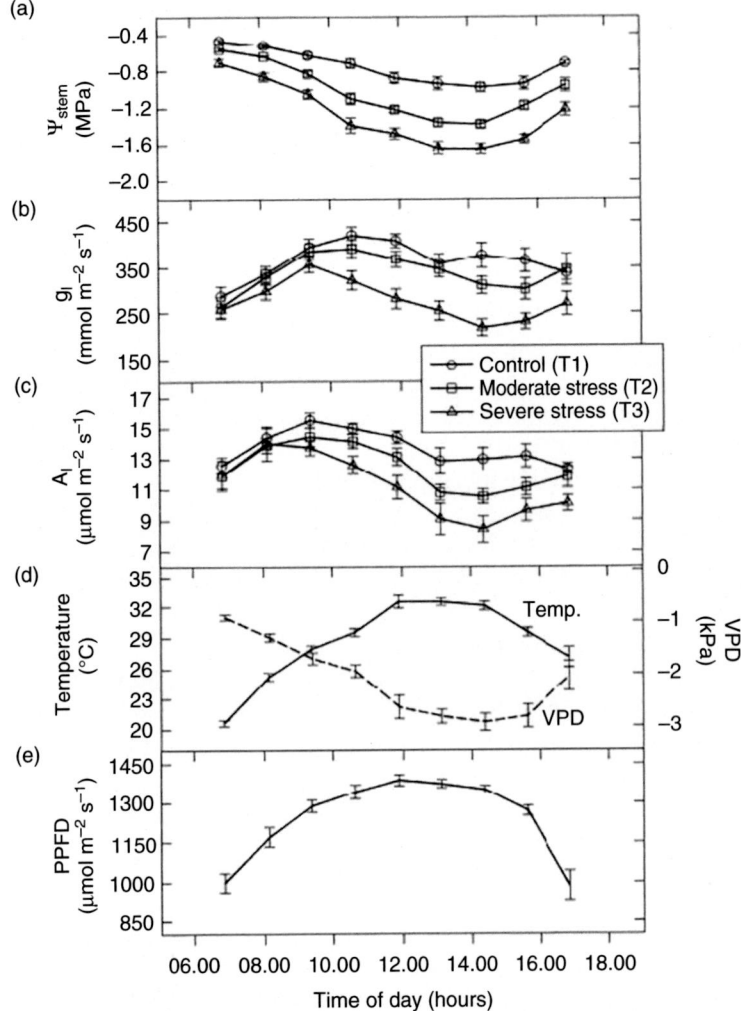

Fig. 2.8. The daily pattern of (a) stem water potential (Ψ_{stem}, an indicator of water stress), (b) leaf stomatal conductance (g_l, stomatal opening), (c) leaf CO_2 assimilation (A_l, photosynthesis), (d) ambient temperature and vapor pressure deficit (VPD, relative air dryness) and (e) ambient light (PPFD, photosynthetic photon flux density) measured on prune leaves in an orchard receiving different irrigation treatments. Symbols indicate mean values and vertical bars indicate standard errors around means. (From Lampinen *et al.*, 2004.)

maintenance functions during the night than during the day. However, this is generally not the case. The cells that carry out photosynthesis in leaves (mesophyll cells) temporarily store enough starch in their vacuoles on a daily basis so that assimilate transport out of the leaves can continue fairly continuously over the whole day and night period. Thus, it is incorrect to think of the

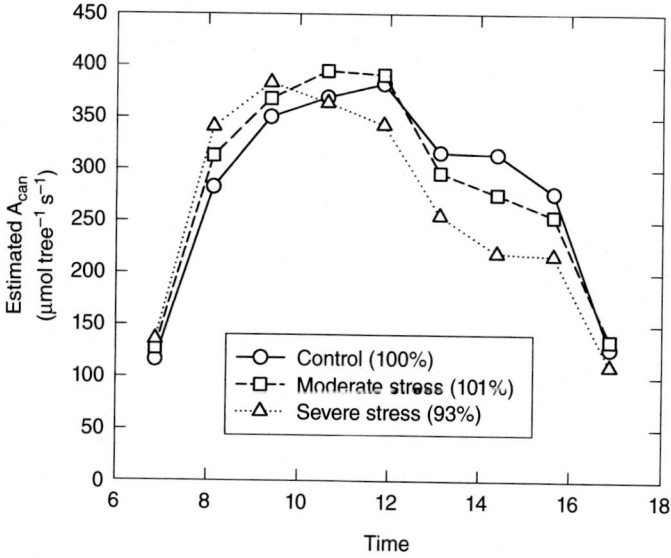

Fig. 2.9. Estimated prune tree-canopy assimilation rate (photosynthesis, A_{can}) over a day of trees receiving different rates of irrigation. Canopy assimilation rates were estimated from measurements of leaf angle and photosynthetic light response curve data. Numbers following treatments in the legend indicate estimated daily canopy assimilation relative to the control irrigation treatment. (From Lampinen *et al.*, 2004.)

metabolism or growth of the tree as "shutting down" during the night because the tree is in the dark. However, there is some day–night alteration of metabolism and growth largely because of diurnal temperature and water relations patterns.

Sink control of photosynthesis

There has been considerable emphasis in the pomological literature on the notion that photosynthesis in fruit trees is strongly controlled by sinks for carbohydrates, especially when fruit are not present or actively growing on trees. Much of the pomological evidence for strong sink control of photosynthesis came from experiments which involved girdled branches or trees growing in pots or on dwarfing rootstocks that artificially limited alternative vegetative sinks and thus caused a sort of "carbohydrate constipation" in the plant such that starch accumulated to high levels in leaves. These kinds of situations are not commonly found in nature or in orchards where dwarfing rootstocks are not used. Furthermore, when considering that ultimately all land plants have been adapted for survival and reproduction, it makes little ecological sense for trees to have a mechanism that limits tree accumulation of photosynthates in the absence of stress under natural growing conditions since carbohydrates are the major currency for sustaining life and competitive ability.

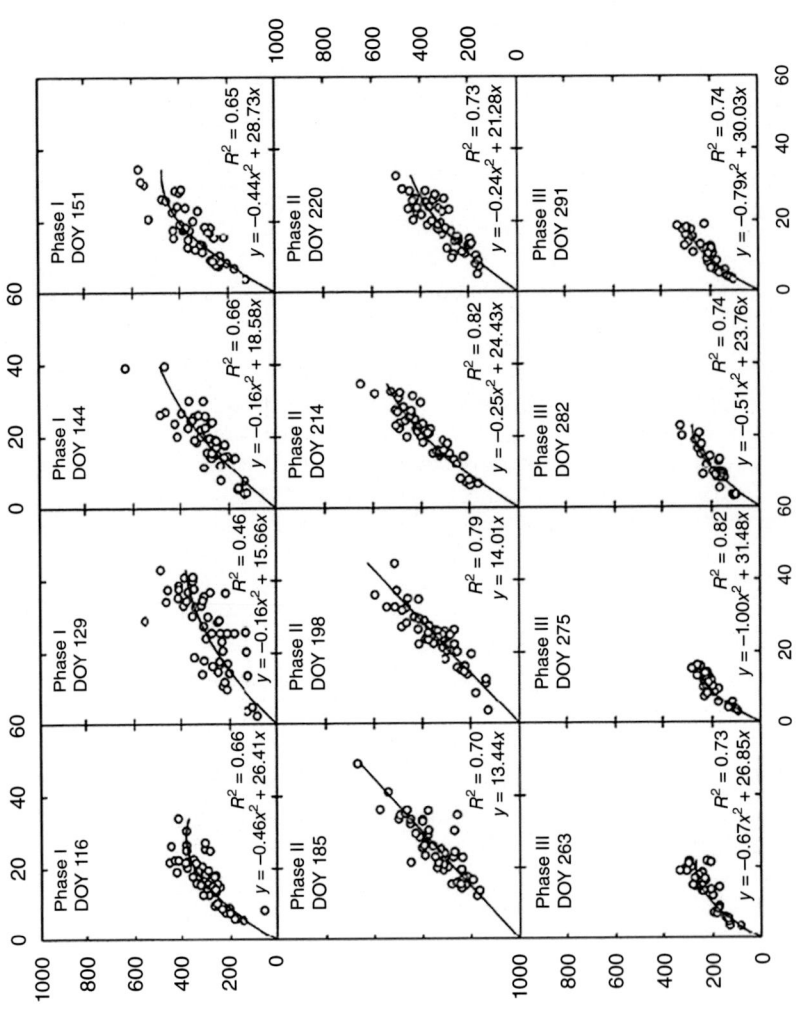

Fig. 2.10. Seasonal changes in estimated daily radiation-use efficiency (photosynthetically active radiation, PAR) and daily canopy photosynthetic net carbon exchange rate (A_n) for apple and pear leaves during early (Phase I), mid (Phase II) and late (Phase III) phases of fruit growth (DOY, day of year). (From Auzmendi *et al.*, 2013.)

However, there is some evidence that the presence of fruit on vigorously growing fruit trees can have a limited effect on stomatal behavior. As stated above, the role of the stomata is to limit water loss while the leaf absorbs CO_2. It has been reported that the presence of actively growing fruit can stimulate stomatal control to be slightly less restrictive to water loss under either no or mild water limitations, so that photosynthesis can be a little higher than if there are no fruit present.

Use of Photosynthates

Ultimately, all photosynthates are used either as building blocks or sources of chemical energy in the synthesis or maintenance of compounds involved in tree metabolism, growth or reproduction. They act as building blocks in the sense that photosynthates are carbohydrates and carbohydrates are the basic "starter compounds" in trees—the compounds from which all other organic compounds are synthesized. The process by which carbohydrates become incorporated into the structural parts of trees is called **assimilation**. Although assimilates start out as carbohydrates, they are used to make widely different compounds such as:

- chlorophyll, carotenes and anthocyanins, pigments which give plant parts their color (green, yellow, red, purple) and are part of light-harvesting mechanisms;
- proteins, which are in the cytoplasm of the cells that carry out the living functions and are constituents of cell membranes;
- nucleic acids, compounds responsible for storing, transmitting and translating the genetic codes for controlling plant construction and function;
- lignin, a class of chemicals that harden the cell walls of wood, bark and pits;
- cellulosic substances, the main constituents of cell walls;
- starch, the main long-term carbohydrate storage compound in most trees;
- oils/fatty acids, high-energy storage compounds that are accumulated by the embryo (seed) during its development and subsequently supply energy for growth during germination, also major constituents of cell membranes; and
- waxes and cuticles, located on the epidermis of fruit, stems and leaves, which prevent water loss and serve as a protective envelope.

Respiration

Carbohydrates also serve as energy sources. Just as a substantial amount of energy is required to synthesize carbohydrates through the process of photosynthesis, the orderly breakdown of those same carbohydrates yields chemical energy for carrying out metabolism in various parts of the tree. Orderly and efficient breakdown of carbohydrates and lipids requires O_2 and is called **respiration**. Effectively, respiration is the reverse of photosynthesis (Fig. 2.3). The substrates for respiration are carbohydrates and O_2, and the products are CO_2, H_2O and chemical energy. Two other major differences between photosynthesis and respiration are: (i) respiration can take place in the dark as well as the light; and (ii) respiration takes place in all living tissue cells and does not require chlorophyll to be present.

Conceptually, respiration is often classified into two types: maintenance respiration and growth respiration. Maintenance respiration is respiration that goes on continually in living tissues to keep them healthy and functioning. All metabolically active compounds, such as enzymes, undergo a continual process of degradation and rebuilding. Thus, a certain basal level of energy is needed to maintain living tissues in a metabolically active state. The amount of maintenance respiration carried on by a tree depends on its physiological state. During the dormant season, when tree metabolism is at a minimum, maintenance respiration is low. But during the active growing season maintenance respiration is higher.

Growth respiration is respiration that occurs to supply energy for the construction of new tissues during the growth of various tree parts. Thus, the growth of trees not only requires photosynthates for building blocks, but also for respiration energy to be used in converting the building blocks into finished products. Even though respiration is the reverse of photosynthesis, it should not be viewed as a wasteful or undesirable process. It is a process essential for tissue growth and functioning. An orchard manager's job is to ensure conditions in which respiration can occur efficiently and wasteful respiration in response to stress is minimized.

Factors Influencing Respiration

Temperature

All living cells respire. The respiration process is catalyzed by several enzymes. These enzymes, being made up of proteins, are temperature sensitive, so that the rate at which they function is temperature dependent. At near-freezing temperatures, respiration rates are low. At temperatures in excess of 104°F (40°C), the enzymes almost become non-functional, so enzymatic rates again decrease. Between 32°F (0°C) and about 90°F (32°C) respiration rates increase exponentially with temperature increases (respiration rate approximately doubles with each 10°C increase in temperature) (Fig. 2.11). This means that on very hot days, whole-tree rates of respiration can often exceed rates of photosynthesis and the net CO_2 uptake of a tree is negative even when other conditions, such as water, are not limiting.

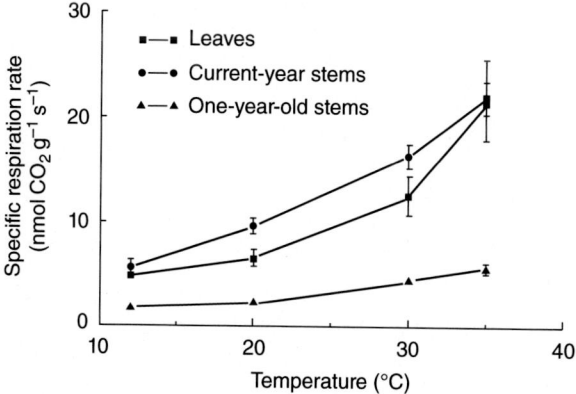

Fig. 2.11. Temperature response of respiration rates of peach leaves, current-year stems and one-year-old stems in May. Note that respiration rates increase exponentially with temperature increases. Symbols indicate mean values and vertical bars indicate standard errors around means. (From Grossman and DeJong, 1994.)

O_2

An essential substrate for respiration, O_2 is rarely limiting to cells of the aboveground parts of trees. However, in heavy clay soils or under waterlogged situations, lack of O_2 often inhibits root respiration and therefore limits root growth and nutrient uptake. It is estimated that, in most years, more orchard trees die from O_2 deprivation in saturated soils than from lack of water.

Soluble carbohydrates

In a healthy, sun-exposed tree, photosynthates are produced in abundance and exported from the leaves to all parts of the plant. At their final destinations they are used for respiration or growth or stored as starch or lipids. Abnormal situations that affect either photosynthesis or translocation of photosynthates—situations such as extended periods of low light intensity, nutrient deficiency or limb girdling—can limit carbohydrate supplied to the cells of specific organs and thus decrease their respiration.

Internal factors

Internal changes in the physiological state of tissues significantly influence respiration rates. Dormant branches in the wintertime respire less than the same branches in the spring, when dormancy is broken. Plant hormones are thought to be active in determining the physiological state of tissues and thus indirectly influence respiration.

Uptake and Assimilation of Nutrient Resources

3

In a manner analogous to the features of aboveground portions of the tree that are specially designed to harvest and capture sunlight, belowground portions of the tree are specially designed to absorb water and mineral nutrients; these are the absorbing roots. Absorbing roots are attached to the tree by larger conducting and skeletal roots that perform the functions of support (anchorage) and conducting connections that allow for extensive exploration of the soil volume to capture water and minerals in a manner analogous to aboveground scaffold branches and their display of leaves for capturing solar energy.

The structure of young, actively absorbing roots is specially designed to filter water and selectively take up needed minerals while excluding many harmful elements. They filter the water by forcing the water and minerals to move through the cells of the endodermis (a layer of cells adjacent to the pericycle surrounding the conducting tissues in the center of the roots). Water moves passively through the endodermal cells while minerals are actively transported across membranes of the cells so that their uptake can be regulated (Fig. 3.1). Water movement from the soil into the plant is primarily driven by a water potential gradient between the soil and the root and the root and the top of the plant. Thus, the root plays a key role in the SPAC.

Many of the minerals are actively taken up into the cortex and/or endodermal cells of the root via metabolically driven transport mechanisms in the cell membranes and these transport processes require energy that is provided through respiration. Root respiration requires carbohydrates that are supplied by the aboveground portion of the tree and O_2 that must be available from the soil. Since respiration is highly responsive to temperature, mineral uptake by roots is also sensitive to soil temperature and the principal reason why most fruit trees grow best on well-drained soils is that O_2 availability to roots limits root respiration when soils are saturated. Both the availability and uptake of minerals by the root are also influenced by acidity (pH), thus soil pH can also have strong effects on the mineral nutrition of trees.

Fruit trees require six macronutrients (N, P, K, calcium (Ca), Mg and sulfur (S)) and eight micronutrients (Zn, Fe, B, Mn, Cu, chlorine (Cl), nickel (Ni) and

© T.M. DeJong 2022. *Concepts for Understanding Fruit Trees* (T.M. DeJong)
DOI: 10.1079/9781800620865.0003

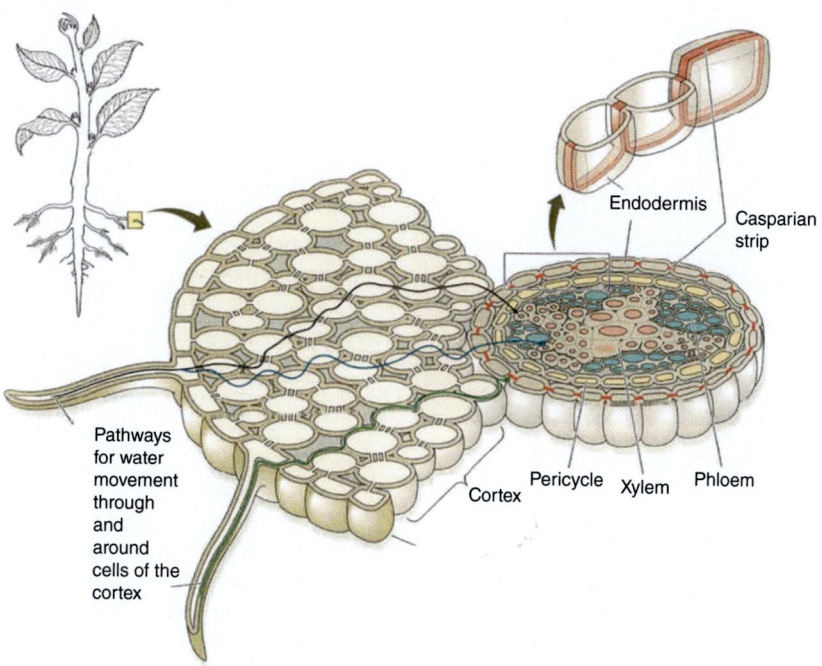

Fig. 3.1. Pathways of water movement into roots. Water and minerals can move either through the cortex cells of the roots or around them until they are forced to move through the endodermis cells of the Casparian strip. (From Taiz *et al.*, 2015.)

molybdenum (Mo)) that are taken up through the roots. Many of these occur naturally in the soil as cations (positively charged particles; NH_4^+, K^+, Ca^{2+}, Mg^{2+}, Zn^{2+}, Fe^{3+}, Mn^{2+}, Cu^{2+}, Ni^{2+}) bound to negatively charged soil particles (Fig. 3.2), while others are dissolved in the liquid surrounding the soil particles in the form of anions (negatively charged particles; NO_3^-, $H_2PO_4^-$, SO_4^{2-}, Cl^-, $H_2BO_3^-$, MoO_4^{2-}). The walls and membranes of root cortex cells and root hairs have complex mechanisms to attract and transport each of these nutrients into the living tissues of the roots so they can be transported around the plant in the xylem and/or phloem. Some of these elements are highly phloem mobile (N, K, P, Cl, S, Mg) and others are less phloem mobile (Zn, Cu, Fe, Mo), while Ca and Mn are almost completely phloem immobile. B is a special case because it can complex with sorbitol in pome and stone fruit trees, and with mannitol in olive trees. These complexes make B phloem mobile in these species whereas it is highly immobile in most other plant species. Relative phloem mobility strongly affects the distribution of nutrients in plants and the more highly phloem mobile a nutrient is, the more likely it will end up in the organs that need it most in a manner similar to carbohydrate distribution around the tree (see the "Distribution of Assimilated Resources in Fruit Trees" section in Chapter 5 of this book). In other words, phloem-mobile nutrients get distributed to the organs that have

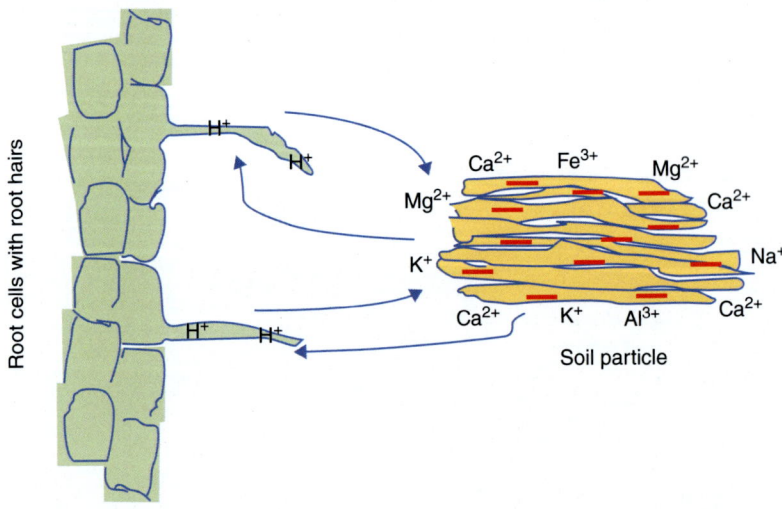

Fig. 3.2. Diagram of the relationship between negatively charged soil particles and positively charged elements (cations) and the exchange of positively charged hydrogen ions for the cations attracted to the soil particles. Thus, the fertility of soils is partially described by their "cation exchange capacity". (Drawing by T.M. DeJong.)

the highest demand for them as they are growing. Deficiency symptoms of less-phloem-mobile nutrients like Zn, Fe, Ca and Mn will often occur in localized regions of the tree rather than more generally throughout the tree, like N deficiency symptoms. Conversely, Zn, Mn and Na toxicity will often occur on leaf margins because leaf margins are the functional end of the transpiration stream in trees and these elements are carried by water flow that is driven by transpiration.

Because trees are perennial plants, storage of nutrients from one year to the next is vitally important to their long-term functioning. It is essential that nutrient resources be available in the spring to supply growth demands as flowering and shoot development resume quickly in the spring while soils are still relatively cold and nutrient uptake is minimal. Thus, significant amounts of nutrients are stored in tissues of perennial organs, particularly roots and branches.

During the growing season substantial amounts of many nutrients can be found in active leaves but approximately half of the amounts of N, P and K that are initially found in active leaves is remobilized out of the leaves and stored in perennial tissues prior to leaf fall (Table 3.1). During the dormant season roots with diameters greater than a pencil account for as much as half of the total stored N and P in a deciduous fruit tree, with the other half of the stored N and P in the upper part of the tree (mainly the branches). Roots are of less importance in the storage of K, S and B in Rosaceous trees and the aboveground perennial organs account for approximately 60 to 75% of their

Table 3.1. Mean annual N, P and K storage (kg ha^{-1}) in perennial organs of mature almond trees that received N fertilizer at 309 kg ha^{-1}. (Data summarized from Muhammad *et al.*, 2020.)

Organ type	N	P	K
Roots < 1 cm diameter	0.15	0.20	0.30
Roots > 1 cm diameter	13.11	2.66	5.77
Trunk	3.65	0.25	1.50
Main scaffolds	4.85	0.90	2.70
Large branches	16.51	2.80	10.28
Branches < 2.5 cm diameter	5.37	1.35	7.34
Total	**43.64**	**8.16**	**27.89**

storage. The concentrations of stored nutrients are often higher in phloem tissue but the total biomass of active xylem tissue is many-fold greater than that of the phloem, so the xylem tissue of roots and branches is the primary storage tissue in a tree.

The Structure of Trees

4

Trees are special types of plants. They use the same or very similar biochemical pathways and cellular processes as other plants and they are subject to the same physical and chemical laws as other land plants, but they have developed different strategies for survival and success than other land plants. All green land plants must have leaves (or similar structures) that capture sunlight and do photosynthesis in order to gain carbohydrates and energy to grow and maintain themselves. And all land plants must be capable of taking advantage of periods when environmental conditions are favorable for photosynthesis and growth, as well as have mechanisms for survival during, or reproduction prior to, periods of stress, such as extreme cold, heat or drought. Annual plants start each growing season from seeds and grow very rapidly, taking advantage of the most favorable periods of the year, and then flower, fruit and disperse seeds for the following years, prior to a period of extreme stress. Biennial plants have mechanisms for growing from seed in one year, carrying out photosynthesis and storing large amounts of carbohydrates that are protected during periods of extreme stress (often in roots or structures in, or very low to, the ground) and then using those stored carbohydrates to grow, flower, fruit and disperse seeds in the following season before they die. The height of both annual and biennial plants is limited because height growth and development of their light-harvesting canopies must occur in the period of one season.

Perennial trees and vines have developed different life strategies. They have developed phenological mechanisms to anticipate periods of stress, stop growth and develop specialized structures (perennating buds) and internal cellular mechanisms and processes that allow their tissues to withstand periods of stress (cold winters, extreme heat, drought), and then resume growth from the points where they stopped prior to the stress. This permits trees and vines to grow tall and maintain their occupation of space over multiple seasons. Deciduous trees can rebuild their photosynthetic canopy very quickly in a growing season by simply growing new leaves on the same framework that was built the previous season. Evergreen trees maintain the same leaves over more than one year but sacrifice some individual leaf photosynthetic efficiency because leaves must be robustly adapted to longer periods of activity. Perennial

DOI: 10.1079/9781800620865.0004

vines are similar in their life strategy to perennial trees except that they do not build their own structure for supporting the plant body but climb and rely on trees or other permanent structures for support. Both trees and vines tend to put greater effort into growing tall and occupying space early in their life history at the expense of delaying flowering, fruiting and seed dispersal until after they are firmly established.

Height or length growth of all land plants is accomplished by cell divisions in apical meristems. Meristems are groups of undifferentiated cells that can divide almost indefinitely. After each division of a meristematic cell, one daughter cell continues to perpetuate the meristem and the other cell, which is proximal to the meristem, enlarges and passes through phases of differentiation until it becomes what it is ultimately destined to be. The process of cell differentiation to form recognizable tissues and functional components of the plant is termed **development**. Plant hormones play key roles as signaling molecules in the coordination of cell differentiation and tissue development.

In shoots, lateral leaf and shoot meristems are formed by divisions of the apical meristem and ultimately form leaves and lateral shoots or buds that facilitate branching of the growing axis (Fig. 4.1). In roots, the apical meristem does not divide or branch, and lateral roots are formed *de novo* by initiation of new lateral root meristems through differentiation of specific cells (in the pericycle) proximal to the apical root meristem. The internal structure of shoots and roots is formed by the differentiation of lines of cells that are produced by the apical meristem as it continues to produce new cells at the apex. This process is essentially the same for annual plants and perennial trees and vines.

Trees are, by definition, the tallest land plants. To grow tall over multiple years they must solve several problems: structural strength; carbohydrate and nutrient storage capacity to survive and regrow after periods of stress; and conductive capacity for water, carbohydrates and nutrients must be increased/renewed over time to keep pace with increases in canopy size. Additionally, apical meristems must be capable of surviving through periods of stress (especially over winter or during drought). Structural strength, storage capacity and water, carbohydrate and nutrient conductive capacity are provided by cells derived from a sheath of meristematic cells (vascular cambium) that surround the body of trees (shoots, stems, branches, trunk, perennial roots). Cambial initial cells divide laterally to provide strength to the major structures of the tree (through fiber cells and lignified cell walls), increase multi-seasonal carbohydrate and nutrient storage capacity of the tree (through xylem and phloem parenchyma cells), and annually renew the vasculature for conducting water and nutrients up the tree (xylem) and carbohydrates and nutrients throughout the tree (phloem) (Fig. 4.2).

Another adaptation for enhancing tree survival over multiple years is the development of a cork cambium (or phellogen). The cork cambium is reconstituted each year on the outside of the phloem (Fig. 4.2) from re-differentiated phloem parenchyma cells to form an outer sheath of meristematic cells that divide laterally to form outer bark (phelloderm) that provides a protective layer on the major structural parts of the tree.

Chapter 4

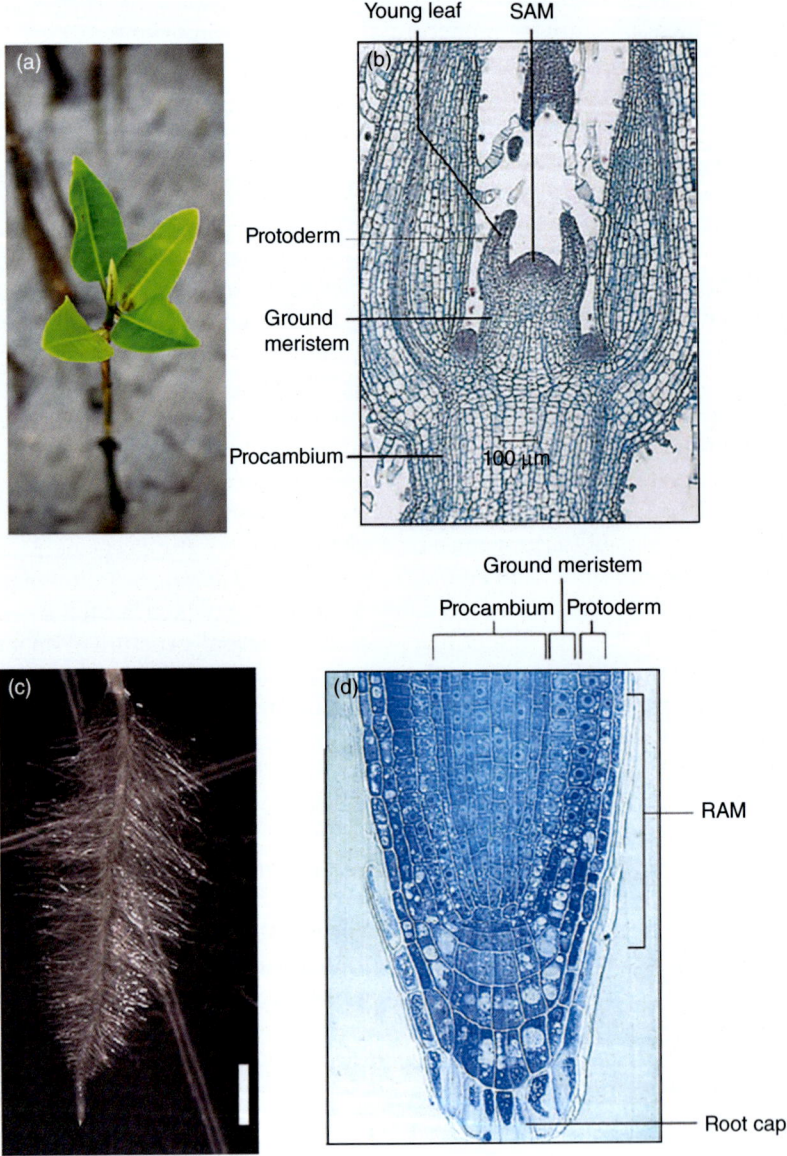

Fig. 4.1. (a, b) Shoot apical meristems (SAMs) and (c, d) root apical meristems (RAMs). Apical meristems provide cells for length growth and the future production of lateral shoots and roots as well as lateral meristematic tissue (cambium) for girth growth of stems. (From Rost *et al.*, 2014.)

A third adaptation is the development of buds containing meristematic apices that have specialized protective coverings (bud scales) as well as cellular mechanisms that resist stress and can go dormant prior to periods of stress (Fig. 4.3). These buds, borne either in the apex of shoots (apical buds) or in the

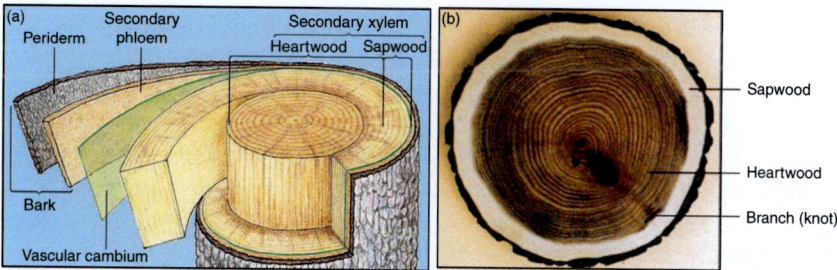

Fig. 4.2. Cross-sectional structure of tree stems. (a) Diagram showing the sapwood, heartwood and other components of a woody stem. (b) Slice of a woody stem of mulberry branch (*Morus* sp.) showing the coloration difference between sapwood and heartwood. The dark area from the center to the lower right is an embedded branch that started to grow when this stem was small. This embedded branch will make a "knot" in a board made from this part of the stem. (From Rost *et al.*, 2014.)

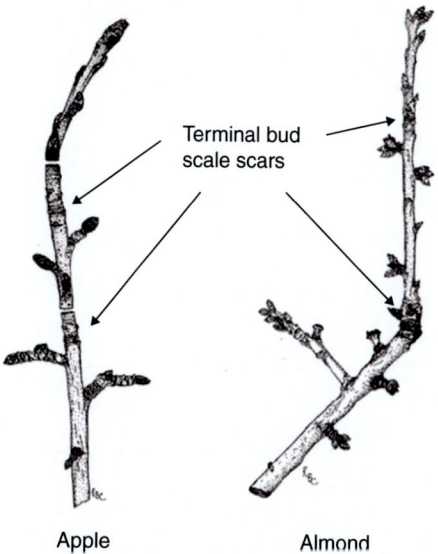

Fig. 4.3. Examples of overwintering twigs of apple and almond showing dormant buds on spurs and terminal bud scale scars that delineate the beginning and end of annual stem growth segments. (Drawings by a student in a Pomology class taught in the 1970s.)

axils of leaves (axillary buds), provide the capacity for apical meristems to go dormant and resume growth after periods of stress. In temperate deciduous trees the primary stress of concern is winter cold, but these buds can also be effective in surviving drought or extreme heat in some species.

These adaptations of trees that are essential for survival and growth over multiple years also provide a practical means for estimating the age of branches or major parts of trees. Tree rings seen in cross-sections of wood when trees or branches are cut down (Fig. 4.2) indicate age because each ring indicates that cambial growth began and then ceased during individual growing seasons.

Similarly, the age of sections of branches can be estimated by counting the bud scale scars from the tip of a branch to its base. At the end of each growing season, a terminal dormant bud has a compact series of bud scales (modified leaves) that leave bud scale scars on the branch (Fig. 4.3), indicating the beginning and end of each growth increment.

The Carbohydrate Economy of Fruit Trees

5

Since all tree growth and maintenance is dependent on the energy and building blocks derived from carbohydrates provided by photosynthesis, carbohydrates are the primary "currency" in trees. Thus, it is important to understand the carbohydrate economy of fruit trees. Figure 5.1 depicts how the dynamics of carbohydrate flow and usage are related to yield and cultural practices applied to fruit trees. Carbon (C) in the form of CO_2 is taken up through photosynthesis and converted into energy-rich carbohydrates that are then translocated to various plant parts. Translocated carbohydrates are used as energy sources for maintenance and growth respiration, or as building blocks (assimilates) in the creation of new plant parts. Note that there are three types of organic-C loss to the tree system. The most important to the fruit grower is the crop that is harvested. Two other important losses are those to pests and diseases and to pruning and litter. Virtually anything the grower does in managing an orchard influences some aspect of the carbon economy of the tree. The goal of efficient management is to maximize photosynthesis and partitioning of photosynthates into economically valuable fruit while minimizing losses to pests, disease or excessive growth that is removed by pruning.

The above- and belowground parts of a fruit tree form a complex, interdependent production system. Roots depend on shoots for carbohydrates and other organic nutrients, and shoots depend on roots for water and mineral nutrients. Hence, any factor that has a direct impact on either of these systems in the short term can indirectly cause changes in the plant as a whole, and in its productivity, over the long term. Water is a particularly good example of a factor that is involved in both short- and long-term processes, because water plays a key role in almost all plant processes. Since lack of water is a commonly occurring condition in nature, plants have developed many physiological responses to help them survive periods of water stress. Most of these responses cause changes in the carbohydrate economy of the tree through reduced photosynthesis, tree growth or cropping, but some of these effects can be managed to have minimal impact on overall tree productivity. Whether these responses influence economic production depends on: (i) the processes occurring

© T.M. DeJong 2022. *Concepts for Understanding Fruit Trees* (T.M. DeJong)
DOI: 10.1079/9781800620865.0005

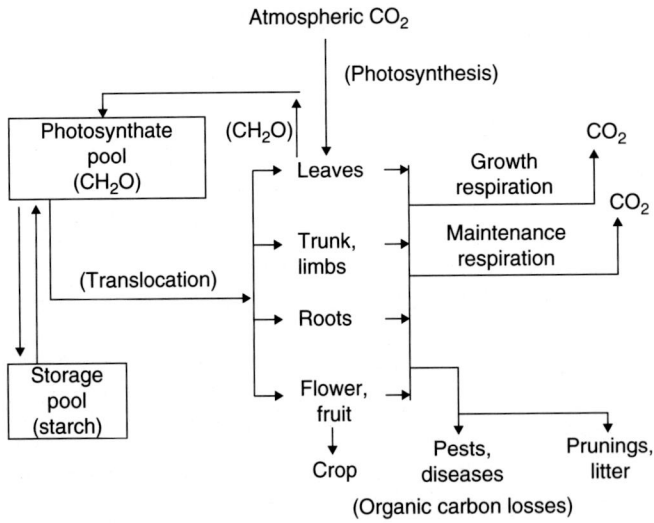

Fig. 5.1. Diagram of conceptualized carbon distribution through a fruit tree. Approximately 50% of the carbon fixed by photosynthesis is used for respiration and a grower's goal is to optimize carbon distribution toward the crop while minimizing losses to pests and diseases and optimizing losses in pruning and litter. (From DeJong, 1989.)

at the time of a stress; (ii) how important these processes are to tree yield; and (iii) whether these processes rely heavily on the current level of photosynthesis or can use stored carbohydrates, like starch, to compensate for the lack of current photosynthesis in the leaves. Not all stresses produce negative outcomes but managing stress to avoid negative outcomes requires increased intensity of management compared with strategies that are designed to routinely minimize stress.

Distribution of Assimilated Resources in Fruit Trees

As mentioned previously, there is a wide range of strategies that plants have adopted to distribute their assimilated resources to achieve capturing resources and producing propagules that allow for survival of the species. On one end of the spectrum there are ephemeral and annual species (such as *Arabidopsis*) that are specialized at sensing when the environment is especially favorable for germination and growth. These plants germinate and grow rapidly when the environment is favorable, then flower and set and disperse seeds within one to several months after germination. Their strategy is focused on rapid growth and reproduction. On the other end of the spectrum are long-lived trees (like *Sequoia*) that germinate their seeds and their life strategy is initially focused on establishment, taking and holding space, and competing with neighboring plants for light and resources over many years. Reproduction takes place only after they are firmly established. These plants have a juvenile period, during which they are incapable of flowering, for one to as many as 50 years, depending on the species and the growing conditions. Temperate deciduous fruit and nut trees are closer to this end of the spectrum. This strategy involves growing a semi-permanent canopy lasting multiple years. Thus, trees must have branches that cease growth and can withstand severe inclement environmental conditions during winter and then resume growth when favorable conditions prevail in spring and summer. This requires them to have special adaptations to protect meristems (tissues of concentrated cell division from which growth originates) during unfavorable periods and sense when conditions will be again favorable for growth.

The structure of rapidly growing annual plants can be determinant and appear to be "preprogrammed" with a specific number of leaves and/or branches being produced prior to nodes that produce flowers or the production of a terminal inflorescence. However, the structure of fruit trees can be highly variable and there is clearly no "preprogrammed" overall structure of a fruit or nut tree that is dictated by the genetic makeup of the tree. Thus, fruit trees can be pruned into many different shapes, and they can even be made up of several different genotypes or species grafted on to a common rootstock. When such a compound tree is constructed, each part of the tree continues to function according to its own genetic make-up (Fig. 5.2). It is very clear that the genetic regulation that occurs within the compound tree is localized within the scions. Thus, the distribution and use of assimilated resources in fruit trees are governed by processes occurring within individual organs of a tree and not by the whole tree.

The following is an outline for how assimilates are distributed and used within a fruit or nut tree.

Assimilate distribution is mainly controlled by the development and growth patterns of individual organs and their ability to compete for carbohydrates and other assimilates:

- A tree is a collection of semi-autonomous organs, and each organ type has an organ-specific developmental pattern and growth potential that are dictated by the genotype of the tree.
- Organ development can be activated by endogenous and/or environmental signals (this could include gene activation, gene transcription, hormones, metabolites, etc.).
- Once activated, environmental conditions and gene actions within the organ determine organ development, which creates conditional organ growth capacity.
- Realized organ growth for a given time interval is a consequence of conditional organ growth capacity (dictated by organ development), resource (carbohydrate and nutrient) availability and inter-organ competition for resources.
- Inter-organ competition for resources is a function of organ location relative to sources and sinks of resources, transport resistances, organ sink efficiency and organ microenvironment.

Fig. 5.2. An example of a compound tree with five scion types grafted on to one rootstock. Note that each scion type maintains the leaf characteristics of each scion genotype. (Photo by T.M. DeJong.)

This concept of resource distribution emphasizes the difference between development and growth. Development involves the differentiation of tissues into specific entities (such as leaves, parts of fruit, components of shoots, etc.) and creates the potential for those entities to grow. Growth fulfills the potential created by development and thus development is a precursor to growth. The genes of a plant contain the blueprints for organ/tissue development, but the organs are only fully built if there are resources available to fulfill their potential for growth. Furthermore, in this scheme there is no overall plan that dictates what or when assimilates go to specific organs. Assimilates produced in the leaves or mobilized from storage tissues are simply loaded into the transport stream (phloem in the leaf during the active growing season or xylem in the woody storage tissues during late winter or early spring) and removed by individual organs as they grow and/or maintain themselves.

This concept of trees being collections of semi-autonomous organs helps in the understanding of how trees develop and grow and what controls growth. The traditional concepts of resource allocation focus on trees as integrated whole organisms that have some centralized control over carbon allocation. These concepts have led researchers to try to identify periods when the "plant prioritized" allocation to specific types of organs at various times during the season or under specific environmental or management conditions. These traditional concepts of resource allocation often led to confusing statements like "the tree allocates more carbon to fruit growth during late stages of fruit expansion". Such statements can't be true because a tree has no mechanism to make such "allocation decisions".

In a way, the decentralized, semi-autonomous organ concept of resource allocation can be thought of as the plant being governed by "democratic capitalism" whereas the traditional view of resource allocation within trees is more like a "centralized planning" sort of government. The semi-autonomous organ concept emphasizes the fact that tree growth can be most easily understood by focusing on the development and growth of key organs that make up a tree. The major periods of growth of organs (and downloading of assimilates in organs) are dictated by the developmental phenology (seasonal patterns of activity) of the organs; the controls for this are mainly in the organs themselves. Once organ development is initiated and an organ begins to grow according to the "blueprint" established by the genetic code for the organ, the structure of the organ is dictated by the genetic code. But the actual size and some characteristics of the organ are dependent on the assimilated resources available to support growth of that specific organ and environmental circumstances that may interfere with the translation of the genetic code. The resources available for growth of each individual organ are dependent on the total assimilate availability in the tree, the competition for those assimilates from other organs and assimilate transport resistances between organs.

(As an aside, this concept of resource partitioning also indicates that traditional views of hormones and plant growth regulators (PGRs) are probably misguided. These compounds are signaling molecules that are thought to convey signals for starting/stopping/enhancing/slowing growth of tissues/ organs. However, if the above concept of resource distribution is correct, then these compounds should be more appropriately referred to as "plant development coordinators" since they primarily coordinate development of tissues and organs, while actual organ growth is dependent on the availability of resources to fulfill the potential growth created by their development.)

Inter-organ competition for assimilates can occur between like organs (such as fruit-to-fruit competition) and between different types of organs (such as shoots, fruits and roots). In naturally adapted fruit tree species, peak periods of growth of major organs often tend to be offset from one another such that the peak period of root growth occurs prior to, or after, shoot growth, and shoot growth generally occurs prior to peak periods of fruit growth, etc. Thus, while there can be significant competition among sinks for assimilates, the developmental patterns of some organs tend to limit competition among the major sinks in trees growing in their natural settings. However, phenological patterns can be adjusted by plant breeding and this can increase competition among different types of organs. For instance, selecting for earlier-ripening fruit tends to increase competition between fruit and shoot growth in peach trees (Fig. 5.3).

Inter-organ competition among organs can also be a function of location of specific organs in the tree relative to other organs. Relative organ location

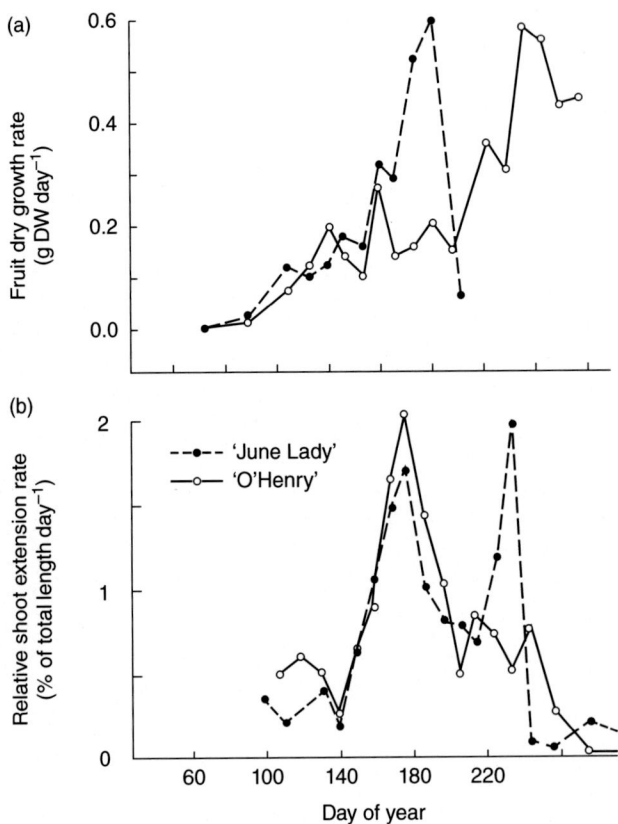

Fig. 5.3. Seasonal patterns of (a) fruit growth rates (DW, dry weight) and (b) corresponding patterns of shoot growth rates for early-maturing ('June Lady') and late-maturing ('O'Henry') peach cultivars. Note that the early cultivar has a second peak of shoot growth after its peak period of fruit growth. This does not happen with the later-ripening cultivar. (From DeJong et al., 1987.)

is especially important because there are significant assimilate transport resistances among various parts of a fruit tree so that some organs naturally have more access to assimilates than others (Fig. 5.4), mainly due to "plumbing". Specific organ location also affects the microenvironment of each organ, and the microenvironment can be quite different for organs in the upper, more exposed portions of a canopy compared with organs in lower, shaded areas of a canopy or in the soil.

To fully appreciate how this scheme functions to control distribution of assimilates in plants it is essential to understand the time-dependent nature of organ growth potential. Organ growth is an iterative process that is dependent on organ development. Organ development occurs over time and determines the state of the organ at specific time intervals and specific growth capacities are associated with each time interval. Organ

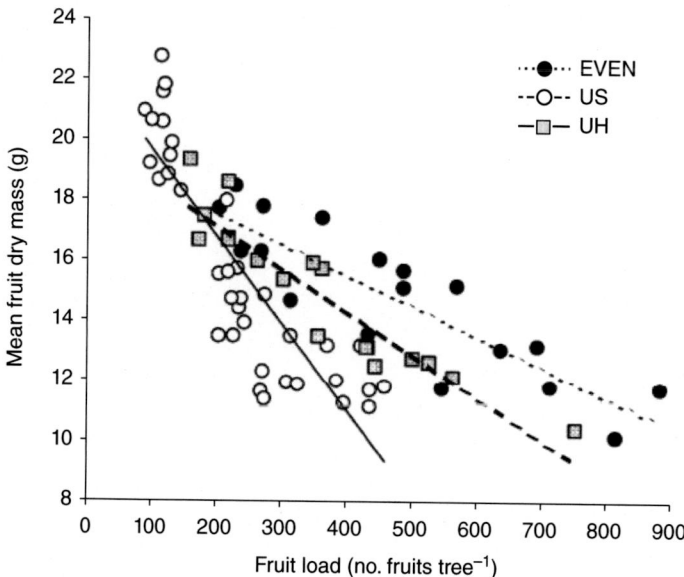

Fig. 5.4. Relationship between mature fruit dry mass and tree crop load on 'Elegant Lady' V-shaped peach trees in which fruit were evenly distributed between scaffolds and among fruiting shoots on the scaffolds (EVEN), only left on one of two scaffolds (US) or unevenly distributed among fruiting shoots on both scaffolds (UH). Data indicate that carbon exchange among different scaffolds and fruiting shoots is very limited. (From Marsal *et al.*, 2003.)

growth cannot occur unless sufficient organ development has occurred, but organ development can occur as long as a minimum threshold of growth has been achieved. If potential growth is not fully realized within a specific time interval, but enough growth has occurred so that development proceeds, then the unrealized growth potential from a previous growth interval cannot be subsequently realized. This concept is common to many biological systems. A good example among humans is that stunted growth early in the development of a child cannot be made up later in life after puberty. Similarly, fruit or leaves in which growth is stunted early in the life of the organ achieve smaller size at maturity but those fruit or leaves still follow through to become mature at nearly the same time as larger fruit or leaves on the same tree.

The functional time interval for realizing the potential growth related to organ development in plant organs is relatively short. Arrested growth over the period of a few days during the growth of a leaf or fruit can influence the size of that leaf or fruit at maturity. Thus, strong competition for assimilates or stress that limits production of assimilates, and thus limits realized growth during relatively short time intervals, can significantly affect leaf or fruit size and yields when the crop is harvested.

To more fully understand how all the major organs of a tree interact in this semi-autonomous scheme of assimilate distribution and tree functioning, it is important to understand their development and growth behavior. The following chapters present a general description of the development and growth characteristics of the major organs of fruit trees.

Understanding the Shoot Sink **6**

Shoot and Leaf Development and Growth

Tree structural growth and architecture is probably most often thought of as something that is arrived at through some centralized control that results in an architecture that is characteristic of a particular species. Indeed, individual tree species or varieties tend to have distinguishable, characteristic canopy structures. However, the resultant structures are more a function of characteristics of individual shoots that make up the tree than a set design for the whole tree. Tree architecture and growth may be more accurately thought of like the construction of a modular building. Each component of such a building has an overarching design but can have owner-specified differences in their actual design, depending on the circumstances. Similarly, tree shoots have an overarching design dictated by genetics but can be modified, depending on their location, environment and timing of growth.

As described previously, all length growth of tree shoots is derived from development and growth of the apical meristem in shoot tips. As the apical meristem grows and divides, it forms leaf primordia and axillary meristems in the axils of the leaves. The location on the shoot where a leaf emerges and an axillary meristem develops is a node. The associated length of shoot between two nodes is an internode, while a node and internode, together, is a metamer or phytomer. Leaf primordia grow into photosynthesizing leaves (solar energy collectors). The axillary and terminal shoot meristems provide the opportunity to continue growth and expansion of the network of solar energy collectors as well as provide locations to later bear flowers, fruits and seeds. As mentioned previously, trees are plants that had to develop specialized structures on their branches to protect apical and lateral meristems from harsh environmental conditions during winter and facilitate the resumption of canopy growth in the spring. Thus, shoots of trees produce apical and lateral vegetative buds that enclose apical and lateral shoot meristems in protective layers of bud scales during the late fall and winter (Fig. 4.3).

© T.M. DeJong 2022. *Concepts for Understanding Fruit Trees* (T.M. DeJong)
DOI: 10.1079/9781800620865.0006

During bud-break, when canopy growth resumes, **proleptic** shoots develop from these previously dormant proleptic buds (Fig. 6.1). Proleptic buds are only capable of producing proleptic shoots in the year after they are formed. Proleptic shoots are the primary shoots that continue the growth of a mature tree canopy. When trees are young and growing vigorously or during other periods of vigorous growth of mature trees, lateral shoot meristems can produce shoots that grow immediately after they are formed in the axils of leaves without a period of dormancy (Fig. 6.2); these are **sylleptic** shoots. Some tree species tend to produce many sylleptic shoots on proleptic shoots while others do so sparingly. Interestingly, nature has also designed trees with the possibility to grow a third category of shoots from preventitious meristems buried beneath the bark at the flanks of axillary buds or shoots (in some species, the location of preventitious meristems can be identified by slightly raised bumps on the lower flanks where stems are inserted into a parent branch). Preventitious meristems keep up with the girth growth of the branch they are on, so they are capable of being activated for many years after a branch is formed (Fig. 6.3). Shoots that grow out of preventitious meristems are called **epicormic** shoots and are often referred to as "water sprouts" or "suckers". Epicormic shoots are nature's way of providing a means for a tree to rapidly replace the volume of canopy lost

Fig. 6.1. An example of a one-year-old peach shoot (reddish-brown) that has produced three proleptic shoots (one terminal and two lateral) from proleptic buds that overwintered from the previous year. (Photo by T.M. DeJong.)

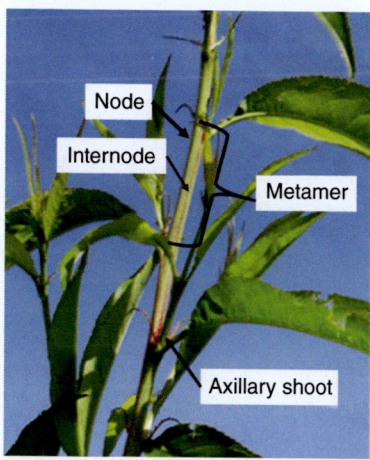

Fig. 6.2. Picture of a peach shoot depicting a node and internode that together make up a metamer. There is an axillary meristem in the axil of every leaf and one of those developed into an axillary sylleptic shoot on this vigorously growing main shoot. (Photo by T.M. DeJong.)

due to major breakage of branches during wind or ice storms. In fruit orchards they are often produced in response to heavy pruning cuts on older branches and can be a major headache for growers who are pruning to control tree size. Since they are naturally produced by the tree to replace what is removed, heavy pruning tends to illicit more epicormic shoots and growers can get into a battle with their trees, especially if they prune heavily during the dormant season (Fig. 6.4). Epicormic shoots often produce numerous sylleptic shoots, depending on the species. When this occurs, epicormic shoots can shade out other fruiting shoots. It is often advisable to prune out most epicormic shoots in late spring after an orchard has been heavily pruned in the previous dormant period.

Overwintering terminal and lateral proleptic buds on temperate deciduous trees usually have about ten preformed metamers and when those metamers expand in the spring, the growth produced is called **preformed** growth. In a mature tree that is unpruned, most of the shoot growth generated in a season is preformed growth. Proleptic buds that produce preformed growth with only minimally expanded internodes result in short shoots called **spurs**. In some species such as almonds, walnuts and apples, spurs are responsible for producing most of the crop in mature trees. However, if a proleptic shoot grows beyond its preformed number of nodes and forms additional nodes, the additional growth is called **neoformed** growth and the shoot exhibits mixed growth. Because sylleptic and epicormic shoots do not arise from proleptic buds with preformed nodes, they are composed entirely of neoformed growth.

The origin and type of shoot growth appear to have some influence on the rate of shoot growth and the length of time that the growth of a shoot remains active. Proleptic shoots that only comprise preformed buds grow very

Fig. 6.3. Cross-section of a three-year-old peach tree branch showing vascular traces connected to two preventitious meristems that grew as the girth of the branch increased after the initial year in which the bud that they were associated with did not develop into a shoot. If the branch had been headed right above the location of these preventitious meristems, it is likely that either one or both these meristems would have produced an epicormic shoot (water sprout). (Photo by T.M. DeJong.)

rapidly in the spring and their extension growth can be completed within a month after bud-break. Proleptic shoots comprising mixed growth usually only grow for two or three months after bud-break and in peach trees rarely attain more than 34 nodes. Interestingly, sylleptic shoots on epicormic shoots of the same peach trees also appear to be limited to a maximum of approximately 34 nodes even though that shoot growth can occur later in the season because those sylleptic shoots can only begin growing after the lateral axillary meristem is formed along the axis of the epicormic shoot. Proleptic shoots of other species such as almond can grow beyond 34 nodes but tend to grow in flushes, indicating there may be some internal regulatory mechanisms that are present

Fig. 6.4. Examples of epicormic peach and plum shoots (water sprouts) that grew in one growing season in response to heading cuts made in the previous dormant season. The shoots were approximately 2 m tall and were still growing in late August. (Photo by T.M. DeJong.)

but can be overridden in vigorously growing shoots (Fig. 6.5). The main axis of epicormic shoots is entirely neoformed and can continue growth through the entire growing season, resulting in production of as many as 80 nodes in peach and other stone fruit.

The structural characteristics of proleptic, sylleptic and epicormic shoots vary greatly among species but relatively few species have been systematically studied and described. For instance, vigorous peach epicormic shoots have a strong tendency to have numerous lateral sylleptic shoots, whereas epicormic shoots of some varieties of almond, prune, apricot and cherry have very few sylleptic shoots (this is a function of genotypic apical dominance characteristics, discussed later). Also, the maximum number of nodes produced in one season on proleptic shoots can vary among species and varieties.

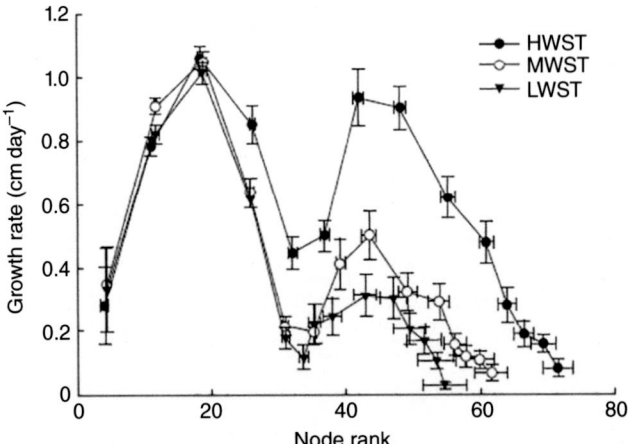

Fig. 6.5. Growth rate of proleptic almond shoots on young trees growing in response to high (HWST), medium (MWST) and low (LWST) water status treatments. Note that in almond the growth rates decrease after approximately 30 nodes are formed but then resume rapid growth for a second flush of growth, and that the second flush of growth is less in trees subjected to the LWST in which there is more water stress. Symbols represent mean values and bars indicate standard errors about the means. (From Negron *et al.*, 2014.)

The rate of shoot extension growth during the season is strongly influenced by temperature, shoot water potential and competition for assimilates. Trees with heavy crop loads usually exhibit less shoot growth. The rate of new node emergence is more rapid in epicormic shoots than proleptic shoots and can also be influenced by temperature and assimilate availability.

Apical Dominance

Terminal and lateral extension development of shoots can be strongly influenced by hormonal control as exhibited by phenomena related to **apical dominance**. In annual plants apical dominance is generally defined as the control exerted by the shoot apex over the outgrowth of lateral buds. However, because of the perennial nature of trees, there is often confusion about the concept of apical dominance as it relates to perennial species. In trees, in addition to suppression of lateral shoot growth in the same season (**correlative inhibition**), there is often control of the length growth of proximal lateral proleptic shoots by the more distal lateral shoots that grow out on the same parent shoot during the subsequent season (Fig. 6.6). The latter phenomenon has been termed **apical control**. Furthermore, there is a third factor that appears to be related to apical dominance and that is **shoot epinasty**. Shoot epinasty is the tendency for the angle of the distal, more vigorously growing lateral shoots on the same parent branch to have narrower

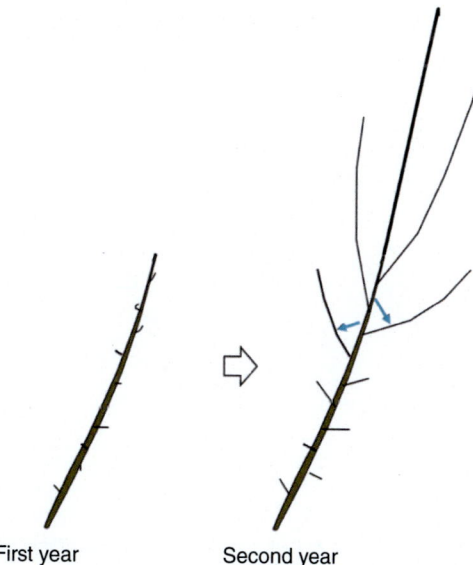

First year   Second year

Fig. 6.6. Diagram of the second-year outgrowth on a first-year proleptic shoot, showing the three manifestations of apical dominance: correlative inhibition (note that none of the proleptic buds grew out in the first year), apical control (note that more distal buds progressively produced longer lateral shoots than more basal buds in the second year) and shoot epinasty (note that the increasing angles of the lateral shoots increase progressively down the stem). (Drawings by T.M. DeJong.)

branch angles than the more proximal shoots (Fig. 6.6). Experiments have shown that the wider branch angle of the lower shoots is related to the same factors that cause apical control as it relates to reductions in shoot length and distance from the shoot tip. The number and length of lateral, proleptic extension shoots that grow out from a previous-year shoot (Fig. 6.6) vary with initial shoot vigor and tree species or cultivar. Some species produce several shoots (as depicted in Fig. 6.6), while others rarely produce more than two extension shoots on a previous-year shoot. Also, some fruit trees often have terminal reproductive buds at the tip of extension shoots (pome fruit and walnut). In these cases, continued growth of the axis of these shoots is dependent on the production of a lateral shoot (Fig 6.7), which is called a bourse shoot in apples.

All three of these phenomena related to apical dominance (correlative inhibition, apical control and shoot epinasty) contribute to what is often described as "acrotony" and appear to be related to auxin transport from the apical growing meristems in distal positions to lateral meristems lower down the shoot as well as other hormonal interactions among leaves and shoots. Since all three of these phenomena (correlative inhibition, apical control and shoot epinasty) appear to be mediated by auxin signaling, they can be thought of as separate but collective manifestations of the general phenomenon of apical dominance.

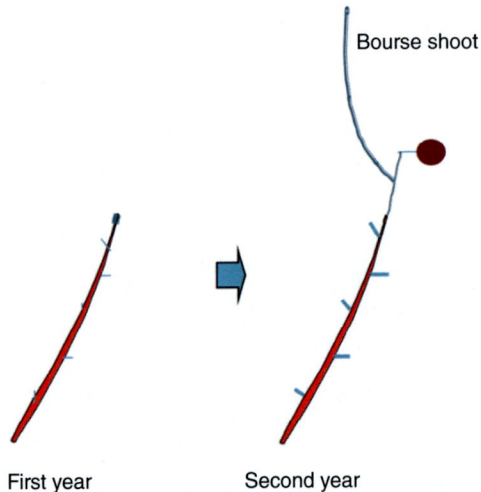

Bourse shoot

First year   Second year

Fig. 6.7. Diagram of the second-year outgrowth (blue) on a first-year proleptic apple shoot. Apical dominance is transferred to the bourse shoot that grew laterally from the short shoot that produced flowers/fruit from a mixed (reproductive and vegetative) bud that grew after bud-break in the second year. Lower proleptic buds on the first-year shoot produced short spurs in the second year. (Drawings by T.M. DeJong.)

Shoot Structure

The anatomical and morphological structure of tree shoots provides for the ability of a shoot to carry out its primary functions: growth in length and diameter; support of leaves and other organs; transport of photoassimilates, water and nutrients to and from leaves; long-term storage of carbohydrates and nutrients; and the initiation of new structures (stems, flowers, fruit and leaves). While shoot length growth and the positioning of leaves, lateral shoot meristems and lateral preventitious meristems are determined by the apical meristem, diameter growth is initiated by the vascular cambium, a contiguous meristematic sheath that is inside the bark. At the center of the vascular cambium are the dividing cells called fusiform and ray initials (Fig. 6.8). These dividing cells determine the internal anatomy of shoots, stems, branches, trunks and woody roots. The fusiform initials produce daughter cells outwardly from the vascular cambium that eventually form sieve cells, sieve tube elements and phloem parenchyma cells. Along with the phloem ray cells produced by the ray initials, these cells comprise the phloem. The phloem is primarily responsible for transporting carbohydrates and nutrients around the plant during periods of active growth and development.

The fusiform initial cells also produce daughter cells inwardly and these cells give rise to tracheids, vessels, fibers and xylem parenchyma cells along with xylem ray initials. Xylem ray initials divide to produce xylem ray parenchyma.

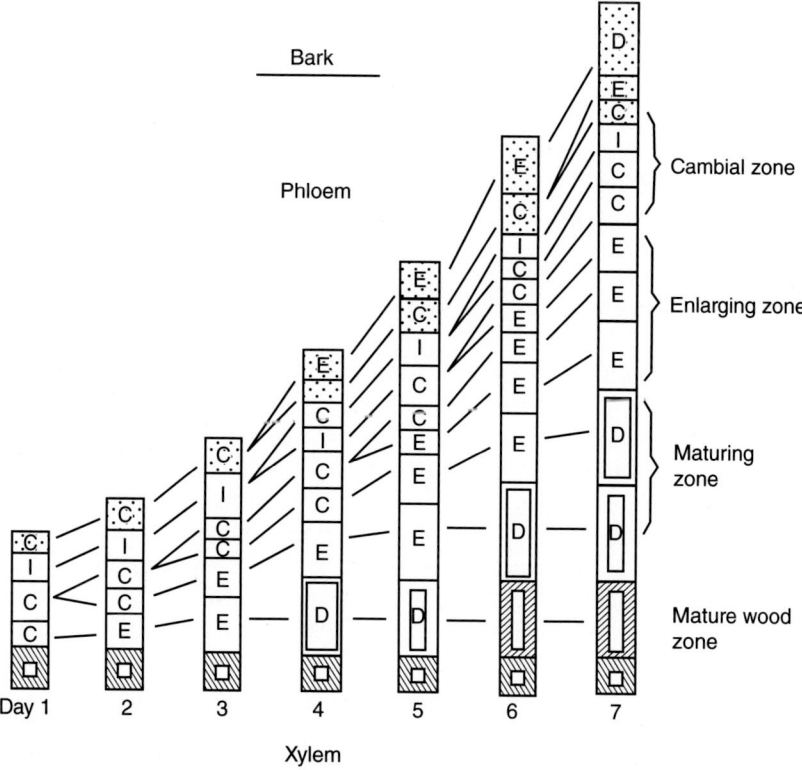

Fig. 6.8. Diagram of the outgrowth of the vascular cambium. The stippled cells form the phloem toward the outside of the tree, the clear cells form the xylem toward the inside of the tree. Cambial initial cells (I) and cambial cells (C) are dividing cells, peripheral to those cells are cells that are enlarging (E) and differentiating (D). Eventually, these cells become mature, functioning xylem and phloem. (From Wilson and Howard, 1968.)

Collectively, the fusiform and ray initials produce the cells that comprise the xylem tissue. The xylem tracheids and vessels are the primary elements for conducting water up the tree from roots to the leaves in the SPAC. The xylem ray cells conduct water, carbohydrates and nutrients laterally across the large roots, trunk and branches.

The xylem and phloem parenchyma cells are collectively the primary long-term carbohydrate and nutrient storage cells in trees. Carbohydrates and nutrients are stored and remobilized from the same xylem parenchyma cells for up to ten years in some species while phloem parenchyma cells are mainly functional for only one year. Thus, while lignified xylem structures provide most of the structural strength of the tree, xylem parenchyma also plays an integral role in the seasonal physiology of trees because it determines the carbohydrate and nutrient storage capacity of the tree.

The functional role of both phloem and xylem in assimilate and nutrient transport is generally limited to less than two years after they are formed in many temperate deciduous tree species. It is also important to recognize that while phloem is the primary pathway for assimilate transport around the tree when the leaves are photosynthetically active, the primary pathway for carbohydrates being mobilized from storage and transported to the top of the tree in the spring is through xylem tracheids and vessels.

Shoot Morphology

The primary morphological feature of shoots is the succession of leaves that form at almost every node along the shoot. At first glance the pattern of structures that appear in the axils of leaves—lateral vegetative buds, lateral floral buds, lateral vegetative buds with associated floral buds, lateral (sylleptic) shoots or no structures (blind nodes)—appears to be randomly distributed. However, systematic analysis of shoots indicates that individual species (or cultivars) tend to exhibit regular patterns of axillary bud fates along shoots and this leads to predictable shoot branching structures. The branching structure of different categories of shoots has been described with statistical models and the patterns of lateral bud fates along individual shoots have been shown to be fairly consistent for specific shoot categories of a given cultivar (Fig. 6.9).

The nodes along a shoot are characterized by the bud fates that occur at each node. Hidden semi-Markov chain analyses of shoots from peach trees (Fig. 6.9) show that shoots have zones or bud fate states that are fairly consistently characterized as containing nodes with specific types of buds. Blind nodes produce no lateral proleptic buds or sylleptic shoots, and in peach are located primarily at the basal and distal ends of the shoots. In peach many nodes produce a lateral proleptic central vegetative bud and this bud can have zero, one or two flower buds associated with it. These types of nodes occupy the middle part of the shoots, with fewer flowers being associated with nodes toward the distal end. Some nodes, in a zone toward the apex of the shoot, produce solitary lateral flower buds. In addition to proleptic buds or sylleptic shoots there are one or two preventitious meristems hidden beneath the bark at each node that maintain the potential to form an epicormic shoot in response to catastrophic injury (or pruning) later in the life of the branch.

As mentioned previously, proleptic shoots can be much longer in almond than peach and growth occurs in flushes. These flushes are apparent in the bud fate patterns of long shoots, with some segments of the shoots appearing to repeat earlier segments (Fig. 6.10). Note that in Fig. 6.10 spurs consist only of preformed growth with ten or fewer preformed nodes and the basal and terminal bud fate patterns are similar to longer shoots. The medium and long shoots are mixed shoots consisting of both preformed and neoformed growth. In peach, proleptic shoots generally exhibit fairly strong correlative inhibition and thus produce few, if any, sylleptic shoots.

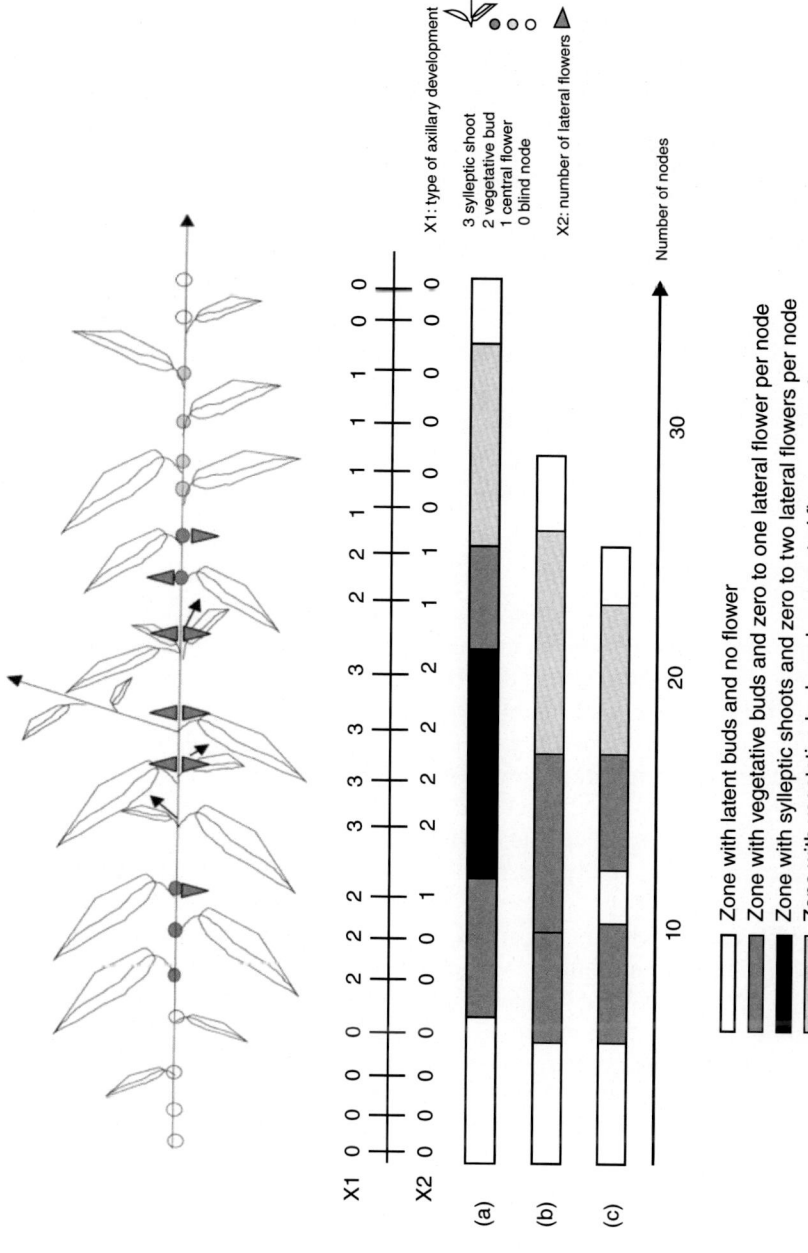

X1: type of axillary development

3 syleptic shoot
2 vegetative bud
1 central flower
0 blind node

X2: number of lateral flowers

Number of nodes

☐ Zone with latent buds and no flower
▨ Zone with vegetative buds and zero to one lateral flower per node
■ Zone with sylleptic shoots and zero to two lateral flowers per node
▦ Zone with vegetative buds and one central flower per node

Fig. 6.9. Diagram showing the bud fate patterns of long (a), medium long (b) and short (c) shoots of peach. Note that long shoots have six bud fate zones and two of those zones are similar, with either blind nodes or nodes with combinations of vegetative and flower buds. Shorter shoots have the same but shorter zones as the longer shoots except that the zone with nodes bearing sylleptic shoots are missing. (From Costes *et al.*, 2006.)

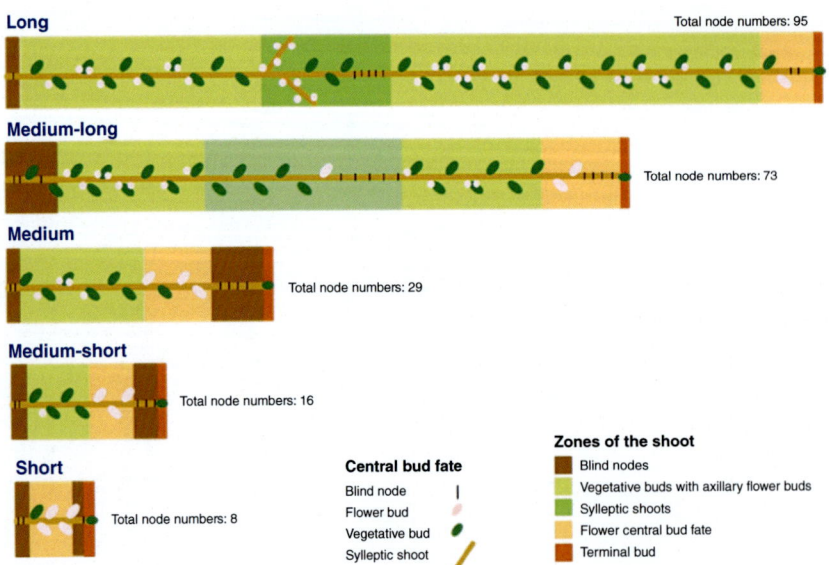

Fig. 6.10. Bud fate patterns of shoots of 'Nonpareil' almond tree. Note that as the length of the shoots decreases, the segments in the mid-section of the shoots disappear but the basal and terminal segments are similar. (Drawings by C. Negron and T.M. DeJong.)

In peach, epicormic (water) shoots are entirely neoformed because the preventitious meristems from which they originate have no preformed nodes. They exhibit very little correlative inhibition and thus produce many sylleptic shoots. Wherever sylleptic shoots are present, there are generally few flower buds produced on the parent shoots and in most cultivars, sylleptic shoots produce relatively few flower buds that set fruit. The main axis of water shoots can have more than 80 nodes produced in one growing season.

The structure and bud fate patterns along shoots can vary widely among fruit tree species and cultivars. However, these patterns are fairly consistent within a species or cultivar but are subject to some variation caused by environmental stresses or horticultural management. For instance, shoot structure and bud fate patterns vary widely among proleptic shoots of different almond cultivars (Fig. 6.11), whereas they tend to be very similar among many peach cultivars. Furthermore, maximum shoot lengths and bud fate patterns among proleptic and sylleptic peach shoots tend to be quite similar despite growing at different times during the season, indicating that the general structure of proleptic and sylleptic shoots may be under similar genetic control.

Shoot Orientation

The orientation of individual shoots on a tree is a function of many factors. The angle of the parent shoot and the angle of bud insertion on the parent shoot

'Nonpareil'

Mean total node number: 95
Mean length: 127.3 cm

'Aldrich'

Mean total node number: 80
Mean length: 124.9 cm

'Winters'

Mean total node number: 100
Mean length: 107.6 cm

Fig. 6.11. Bud fate patterns of long proleptic shoots of three almond cultivars. Symbols and colors represent the same things as in Fig. 6.10. Note that both 'Nonpareil' and 'Aldrich' tend to produce sylleptic shoots only in the middle of the shoots (corresponding to the beginning of the second flush of growth) while 'Winters' produces many more sylleptic shoots along the length of the proleptic shoot. (After Negron *et al.*, 2013.)

(a function of shoot epinasty) are both important for determining the initial direction of growth. But the subsequent direction of shoot growth is often also affected by phototropism and geotropism. It is well recognized that shoots tend to grow toward areas with more available light and this certainly can play a role in the directional growth of shoots, but it may not be of major importance in orchard situations where trees are spaced to receive adequate light. Some of the apparent phototropic effects on shoot growth of orchard trees are probably related to the fact that shoots receiving more light naturally grow more because they have more photosynthetic resources than shaded shoots. But this is not considered a true phototropic response.

Geotropism (growth in response to gravity) is common to almost all fruit trees and strength of the geotropic response has a large influence on whether shoot growth tends to be more upright, as with pruned sweet cherry or pear trees, or more spreading, as in some almond and apricot cultivars. As described above, shoot epinasty (an expression of apical dominance) can also have a substantial effect on branch orientation. The widening of branch angles of proleptic shoots progressively down a branch (Fig. 6.6) is likely a collective response to hormonal signals generated by the shoot apex and lateral shoots proximal to a shoot.

As branches age and must bear increasing weight, the angle of branches tends to droop because of increased weight but this is naturally counteracted by "reaction wood". Reaction wood is the stimulation of asymmetric growth of structural xylem tissue on the upper side of branches (tension wood in fruit trees) to reinforce their ability to carry increasing weight as branch length and cropping increase.

Diurnal Patterns of Shoot Growth

Under normal circumstances, the daily period of maximum shoot extension growth in fruit trees occurs in the mid-to-late afternoon hours. This is counterintuitive because it is well known that individual plant cell expansive growth depends on cell turgor pressure, which is a function of cell/plant water potential. Thus, it is logical to think that, since tree water potential is highest (least negative) during the predawn hours of the day, the most rapid shoot extension growth should also occur in the predawn or early morning period. The reason why maximum shoot extension growth occurs in the mid–late afternoon is that temperatures tend to be highest at that time and tree water potential begins to rapidly recover from a mid-day minimum in the afternoon.

Shoot water potential decreases during the day as leaf transpiration increases until mid-day because the water flow through the soil to roots and through the tree cannot keep up with canopy transpiration. When shoot water potential decreases until mid-day and daily photoassimilates accumulate, the cells in the shoots take up more solutes that attract more water and this helps to minimize the deficit in mid-day water potentials. Then, when canopy transpiration rates decline after mid-afternoon due to partial stomatal closure and decreasing temperature and solar radiation, plant water potential increases and water becomes relatively more available in the shoot cells. This leads to an increase in cell turgor pressure that stimulates rapid cell expansion. Thus, daily shoot extension growth is somewhat analogous to blowing up a balloon. A balloon expands primarily during the period when increased pressure is exerted on the balloon walls but not during the period when the pressure is steady.

For the above reasons shoot and fruit expansion growth are very strongly affected by periods of tree water stress. Thus, it is important to maintain proper irrigation to maximize growth, especially when trees are young and growers are interested in rapid canopy growth. This is also the reason why it is possible to partially control shoot growth with periods of moderate water stress when fruit growth is not a factor. However, it is virtually impossible to simultaneously control shoot growth with mild water stress and maintain maximum fruit expansive growth in most fruit species. Thus, control of shoot growth by imposition of water stress is not practically feasible prior to harvest of many fleshy fruit crops.

Seasonal Growth Patterns of Shoot Growth and Dormancy

In virtually all temperate deciduous tree species, the primary period of rapid growth occurs in the spring as daily temperatures increase and danger of hard, extended freezing weather is past. The period from two to eight weeks after bloom is sometimes called the "grand period" of tree growth. During this period the preformed internodes in proleptic buds expand like a telescope and, subsequently, additional nodes develop resulting in neoformed growth and mixed shoots. Similarly, during this period, growth of preventitious meristems (buried under the bark lateral to the location of buds or shoots from previous years) can be stimulated to form epicormic shoots (water shoots) especially below pruning cuts on branches greater than one year old.

The growth of proleptic shoots of many fruit tree species ceases after a couple of weeks to three months depending on how many nodes are added to the shoot. This means that the growth of most proleptic shoots in many species ceases by mid-June if not earlier. The reason for this does not appear to be environmental because, in the absence of applied stress, conditions for shoot growth in June are generally very good. It is interesting that in peach trees both proleptic and sylleptic shoots are rarely more than 34 nodes long, indicating that their maximum node number is under genetic control. On the other hand, epicormic shoots can continue to grow throughout the summer and into early fall, if environmental conditions are favorable. The termination of growth in epicormic shoots appears to be primarily determined by ambient environmental conditions (decreasing day length and lower night temperatures).

Shoot growth rates are sometimes measured in terms of the rate of addition of nodes (or leaves). The amount of time elapsed between the appearance of subsequent leaves on a shoot is termed the phyllochron. In many annual crops it is thought that temperature is the main driver of the phyllochron and it decreases (nodes are added more quickly) with increasing temperature until a maximum temperature threshold is reached. However, the phyllochron of field-grown peach trees varies depending on the vigor of the shoot and the period of the season (the time interval between the addition of new leaves increases as the season progresses in spring and early summer), and can vary with crop load. In general, the phyllochron for field-grown peach trees during spring and early summer in California ranges between two and three days. Data on other tree species are very limited.

Growth in tree girth is also initiated in the spring as the vascular cambium becomes active, generating new phloem tissue to the outside and a new ring of xylem tissue toward the inside (Fig. 6.8). In actively growing trees, the period of active girth growth can extend until fall and the rate of girth increase can be nearly linear over most of the growing season if the tree is not stressed. However, there is a distinct diurnal pattern of expansion and contraction in girth when measured hourly. Girth expansion is also very susceptible to diurnal patterns of tree water potential because collectively the xylem tissue of a tree represents a sizable reservoir for water storage. Thus, girth growth is very susceptible to tree water stress. Seasonal girth growth can also be inhibited by crop load (Fig. 6.12).

On a seasonal basis the growth periods of a tree are primarily regulated by patterns of **dormancy**. Conceptually, it is easiest to think of the normal state of a plant as having development and growth of its active meristems being "turned on". Dormancy is the state that plant tissues are in when they are "turned off" or metabolism is substantially reduced. So, dormancy is the state of a plant tissue when that tissue is not rapidly growing. Thus, during dormancy very

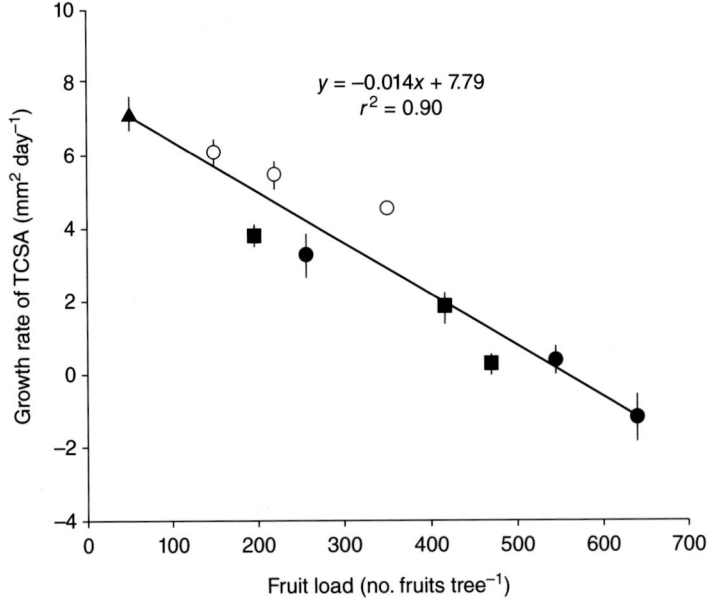

Fig. 6.12. Relationship between growth rate of trunk cross-sectional area (TCSA) and tree crop load in treatments with different crop loads. Growth in trunk girth strongly decreased with increasing crop loads. Symbols represent mean values and vertical bars indicate standard errors. (From Marsal *et al.*, 2003.)

limited cell division or expansion occurs. Dormancy is categorized according to three mechanisms thought to be involved in imposing dormancy.

Paradormancy is when the development and growth of meristems is arrested by signals (mainly hormonal) coming from adjacent plant tissues or organs. The best example of this is with correlative inhibition (a manifestation of apical dominance) in shoots where actively growing apical meristems and the presence of leaves play a role in imposing dormancy on lateral or terminal meristems (buds) in proximal positions on a shoot. This condition occurs in summer when environmental conditions are conducive for growth but growth is held in check by other active organs.

Ecodormancy occurs when growth and development are arrested by environmental conditions surrounding the meristem or tissue. This can occur when temperatures are too hot or cold to be conducive for growth or under conditions of severe water or nutrient stress. This can also occur if plant tissues become starved for assimilates due to excess shading or any number of environmental phenomena. If an imposed stress is relieved and environmental conditions are otherwise conducive for growth or metabolic activity, then growth or metabolic activity can resume.

Endodormancy (sometimes called **rest** or **true dormancy**) is when growth is arrested by factors within the meristem (bud) or tissue itself. This is the case for overwintering buds or meristems in temperate deciduous fruit trees. Although, for all practical purposes, it is assumed that growth and development are completely stopped during endodormancy, careful examination of dormant buds indicates that cell division and organ development do occur even during dormancy but at a much reduced pace.

Endodormancy can be thought of as a mechanism to prepare buds or meristems to survive winter and to prevent them from reactivating rapid growth prior to a period when weather conditions are "safe" to resume growth. Actively growing tissues and dividing meristems are particularly susceptible to freeze damage. When buds go into endodormancy, cell division, and especially cell expansion, occurs very slowly and tissues undergo "winter hardening" to aid in withstanding cold weather with minimal tissue damage. It is important that tissues don't "deharden" before danger of severe winter weather is past. To achieve this, shoot and flower buds of temperate deciduous trees must be exposed to an extended period of **winter chilling** before buds return to ecodormancy.

The amount of chilling required by specific fruit tree species or cultivars depends on the environment that they were native to and any breeding for altering the chilling requirement that occurred during their domestication. Thus, species native to more northern latitudes or cold mountain valleys, like cherry and apricot, tend to have higher chilling requirements than species like almond that originated in more moderate temperate climates. However, there has been much activity in breeding for "low-chill" cultivars in many species so there is a wide range of chilling requirements among currently available cultivars of many species.

The **chilling requirement** of specific cultivars has traditionally been measured as an accumulation of the number of hours below 45°F (7°C) or between 32°F (0°C) and 45°F (7°C). Recently there have been several chill models that give more weight to specific temperature exposures or try to account for the effect of intermittent periods of cold and warm temperatures that can occur during a specific winter period (see http://fruitsandnuts.ucdavis.edu/Weather_Services/chilling_accumulation_models, accessed 18 August 2021). After a specific fruit tree cultivar is exposed to sufficient winter chilling, the buds on the tree go into ecodormancy and exposure to warm spring weather stimulates bud-break and the resumption of rapid development and growth. The amount of heat required to stimulate regrowth is somewhat dependent on previous chilling exposure. If chill exposure is more than adequate, the amount of warm temperature exposure required to stimulate bud-break and regrowth is minimal and the bloom period tends to be compact (occurs over a short period of time). If winter exposure to chilling temperatures is marginal, longer exposure to warm weather is required to stimulate bud-break and regrowth. Under these circumstances bud-break/bloom can appear erratic and be extended over a longer period than normal. If chilling is inadequate, bud abortion may occur and many of the nodes along shoots may remain "blind" (proleptic bud abortion), resulting in two-year-old stems with sparse spurs and lateral shoots (Fig. 6.13).

Fig. 6.13. Spring growth on previous-year shoots of a peach tree that was exposed to a lack of chilling in the preceding winter. Note the lack of leaves in the mid-section of the shoots. (Photo by G. Reighard.)

The mechanism that a tree uses to measure the amount of chilling it is exposed to during winter months has been a mystery for many years. Recently it has been discovered that, while most winter carbohydrate storage occurs in parenchyma cells in the form of starch (which is inert), parenchyma cells in stems maintain fairly constant concentrations of soluble carbohydrates (sugars and sugar alcohols). To do this requires active regulation of starch synthesis and degradation and modulating the activity of enzymes involved in these processes in response to daily temperature fluctuations. Starch synthesis enzymes are more sensitive to temperature than starch degradation enzymes so maintaining relatively constant soluble carbohydrate concentrations over varying day/night temperatures requires constant metabolic adjustments. In spring, as prevailing temperatures increase, since starch degradation enzymes are less sensitive to temperature than starch synthesis enzymes, relatively constant soluble carbohydrate concentrations cannot be maintained. Therefore, they decline, and this is thought to trigger bud-break. This theory is still being fully developed but holds great promise for finally solving the mysteries behind chilling requirements.

Tree Aging

When thinking about patterns of tree aging, it is useful to be mindful that the life strategy of trees is to germinate and initially be directed toward establishment and successfully competing for light and nutrients, before producing reproductive propagules for subsequent generations. Thus, after seed germination, in their natural setting trees go through a juvenile, sapling phase during which they neither flower nor fruit. Subsequently they become mature and are capable of flowering and fruiting.

However, in an orchard situation virtually all commercial fruit and nut tree species do not start out from seedlings. Instead, commercial fruit trees are generally produced from buds or grafting wood taken from mature trees (capable of flowering and fruiting) that is grafted on to rootstocks or are produced from mature rooted cuttings. Thus, an orchard manager almost never deals with truly juvenile trees.

However, tree juvenility is important to the production of pomological crops, in general, because of its effect on fruit tree breeding. Seedlings of fruit tree species can have juvenile periods ranging from one to ten or more years. The length of the juvenile period can greatly decrease the efficiency of breeding new cultivars of a given fruit tree species. If a species had an average juvenile period of five years, it would take approximately 30 years to complete six generations of breeding (with annual crops, it is not uncommon for plant breeders to accomplish 60 generations of crosses in the same amount of time by growing seedlings, in both the northern and southern hemispheres). Thus, tree crops are generally much less genetically developed than annual crop species.

Knowledge of the characteristics of juvenile species can also be helpful in identifying rogue seedling rootstocks. Juvenile shoots often have thorns or spines. Also, vegetative propagation of some species can be facilitated by using shoots from juvenile trees rather than using shoots from mature trees.

After orchard trees reach their optimal period for fruit production they usually begin to decline and gradually go into senility. The length of the prime productive period for orchard trees is dependent on the species, the environment and orchard management. Some species such as peach and nectarine can live for up to 50 years but their prime productive period is usually less than 20 years, while other species such as pear can be highly productive for over 100 years. Tree stress from biotic and abiotic factors certainly plays a role in the length of the productive period but there also appear to be significant differences among species in this regard.

Application of Shoot Growth Rules for Understanding Responses to Pruning

Fruit tree pruning is done to develop vigorous, mechanically strong frameworks, obtain well-shaped trees for convenience of orchard management, achieve optimal capture and distribution of sunlight throughout the canopy, promote fruit size (crop thinning technique), stimulate new fruiting wood (bearing wood renewal), regulate crop production over years (decrease alternate bearing) and control tree size. The last objective is the most common objective that comes to mind but is the most difficult to achieve under good growing conditions. It is natural to think that heavy pruning can be used to reduce tree size.

However, knowledge of fruit tree shoot types is helpful to explain why pruning is often not successful in reducing tree size. Epicormic shoots in fruit trees are primarily generated from preventitious meristems in response to branch breakage, severe bending of limbs (generally below horizontal) and severe pruning. In many horticultural circumstances, epicormic shoot growth can be considered as being almost exclusively stimulated by severe pruning of large branches (older than one year old) or strong water shoots in which sylleptic shoots have previously grown and "used up" the locations in close proximity to the pruning cut where proleptic buds would have been present in a less vigorous shoot. The strong growth response to heavy pruning (Fig. 6.4) is natural and is the primary reason why pruning cannot be relied upon exclusively to control tree size when trees are grown in highly fertile soils without size-controlling rootstocks.

If proleptic shoots that are less than one year old are headed, they are usually replaced by a shoot(s) of the same category arising from lateral buds in close proximity to the cut (Fig. 7.1). If no pruning cut occurs on a given shoot, new proleptic shoots arising near the apex of a shoot are generally either the same or smaller than the parent shoot and lateral proleptic shoots are progressively smaller further down the shoot because of apical control effects (Fig. 6.6). Thus, pruning less vigorous proleptic shoots results in some regrowth but the regrowth is not as vigorous as cutting older branches or vigorous epicormic shoots.

Making thinning cuts (removing stems that grow laterally off the main axis of a branch) will almost always elicit less regrowth than making heading

© T.M. DeJong 2022. *Concepts for Understanding Fruit Trees* (T.M. DeJong)
DOI: 10.1079/9781800620865.0007

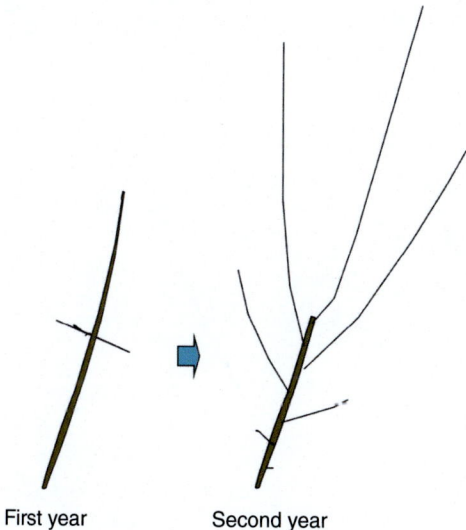

First year   Second year

Fig. 7.1. Diagram of shoot growth response to a heading cut on a proleptic shoot. Note that all aspects of apical dominance (correlative inhibition, apical control and shoot epinasty) are re-established in new shoots below the cut. (Drawing by T.M. DeJong.)

cuts (removing the tip region off a branch) (Fig. 7.2). This is because removing the end of a branch removes some of the natural growth-controlling mechanism of the branch (sources of apical dominance effects) and it alters the ratio of the remaining potential growing points to remaining biomass. Lateral buds tend to be closer together at the distal ends of shoots compared to the proximal ends of a stem while stem diameter is greater at its base than at its tip. Making thinning cuts removes entire shoots so that the balance between biomass and potential growing points is less disturbed.

Pruning can also have strong effects on branch angles. Lateral branches on young, unpruned cherry and pear trees are often oriented at wide angles from the parent central shoot due to shoot epinasty. If one does heavy pruning and heads the central leader, almost all subsequent vigorous lateral shoots will be very vertical with narrow crotch angles. Once these new shoots are established, repeated pruning in attempts to correct it often results in subsequent growth with similar upright growth characteristics. This phenomenon has led to development of new pruning systems for these types of species that emphasize trying to reinforce the natural tendencies of tree growth rather than trying to enforce a specific structure through heavy pruning.

In California, as well as in many other fruit production areas, most fruit trees are propagated by bud-grafting scion cultivars on to newly sprouted rootstocks in late spring. The buds used for this are excised from new shoots produced the same spring and thus they are not fully developed proleptic buds. Therefore, the shoot that grows after bud-grafting is totally neoformed and

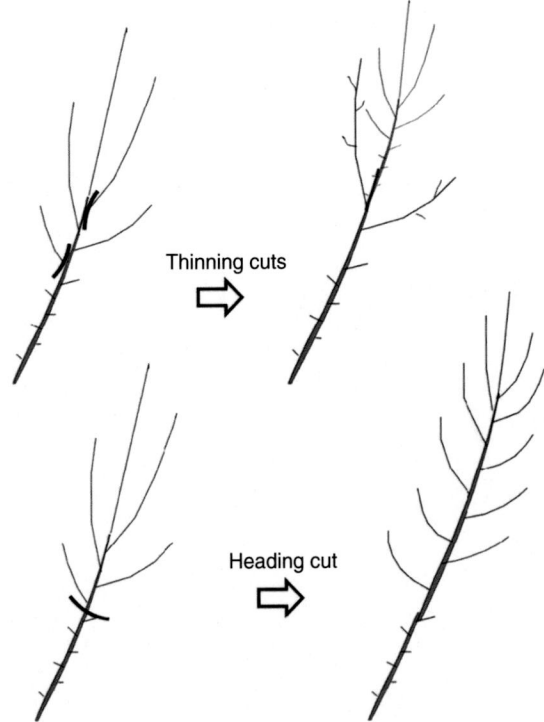

Fig. 7.2. Diagrams of pruning response to two types of pruning cuts on the same two-year-old peach tree branch. With thinning cuts, entire shoots are removed but others are left intact. New proleptic shoots grow on the previous year's shoots. The heading cut into the two-year-old branch stimulates the production of an epicormic shoot and note that many of the sylleptic shoots on the epicormic shoot grow to similar lengths and do not exhibit apical control or much shoot epinasty. (Drawings by T.M. DeJong.)

behaves like an epicormic or water shoot. In peach this shoot produces many sylleptic branches, but in almond it can be a single long whip. Because these are functionally epicormic shoots they have an extended period of growth into the fall resulting in a tree that can be 1.5–2.5 m tall with as many as 70 nodes after one season in the nursery (Fig. 7.3, year 1).

When these trees are removed from the nursery and planted in a production orchard, they are generally headed to a height of 0.5–1.0 m above the soil (Fig. 7.3, year 2). This heading cut stimulates a reiteration response and several shoots are initiated below the pruning cut. In peach these new shoots originate from preventitious buds and thus are epicormic, entirely neoformed and grow late into the season. The classic "open vase" training system used in California involves selecting three or four of these vigorous water shoots at the end of the first year in the orchard to establish main scaffolds and heading them again, so they are ~0.5 to 1.0 m in length. (The "perpendicular V system" involves selecting two of these vigorous water shoots and heading them again as described in the "open vase" system (Fig. 7.3, year 3).) If the heading cuts on the selected scaffold branches are in the vicinity of sylleptic shoots produced the previous summer (as in "short pruning"), they again stimulate production of new water sprouts. If they are headed near the tip of the branch (as in "long pruning"), the reaction to pruning will be less and new lateral shoots the following season will be from proleptic buds and thus less vigorous. With the classic vase system the "short pruning" is repeated for a third year to establish a set of two tertiary scaffolds on the top of each secondary scaffold. At the end of the third year this classic open vase tree has a strong structure. However, the water sprouts produced as a result of the previous year's heading cuts are so tall that the grower is compelled to prune them fairly hard again because, without pruning, they will begin to bend over with crop and/or the tree is taller

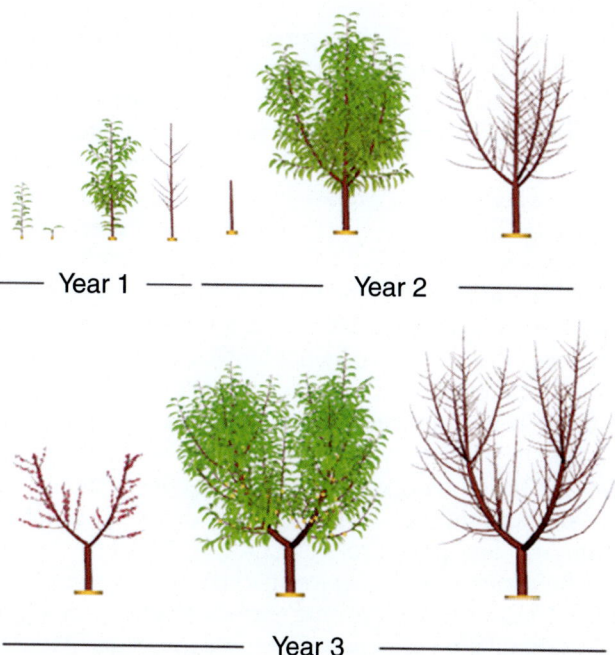

Fig. 7.3. Simulated growth of a two-leader peach tree over three years. In the first year the tree would be growing in a nursery. In the second the tree would be planted in an orchard and headed at about 0.5 m and all the side shoots removed. The hard heading cut stimulates strong epicormic shoots. In the third year two of the epicormic shoots are selected for scaffolds and headed again. Although not shown, after the third year of growth one shoot on each of the scaffold branches would be selected to extend its length. Depending on the preference of the grower, either additional heading cuts (to stimulate more length growth) or thinning cuts (to slow down their length growth) will be applied. (Computer graphics generated using L-Peach.)

than is optimal for management. In vigorous growing conditions this results in a non-productive cycle of pruning and excessive growth responses.

In many species like almond, prune, plum, apricot, cherry, apple and pear, many cultivars produce epicormic shoots with very strong correlative inhibition (the opposite of peach and nectarine) and thus have very few, if any, lateral sylleptic shoots. Vigorous shoots of these species also often exhibit very upright growth. Responses to pruning of these species are generally commensurate with how strongly they are pruned. Hard pruning stimulates the regrowth of strong shoots and the extent of sylleptic branching varies with the correlative inhibition characteristics of the cultivar/species (cherries often produce strong, vertical growth with few sylleptic shoots whereas almonds and plums can produce modest sylleptic branching). Proleptic spurs and lateral proleptic shoots are produced in the following year. Thus, less severe pruning (long pruning) is often recommended for these species after the primary scaffolds have been selected.

Traditionally, most pruning of temperate deciduous fruit trees has been done in the dormant season. However, it generally does not harm a temperate deciduous fruit tree to be pruned during almost any period of the year unless stem-infecting diseases are prevalent in a region. If stem-infecting diseases are prevalent in a region, it is often advisable to do most pruning during drier periods of the year when pruning wounds can dry out more quickly and resist infection (e.g. pruning apricot and cherry in the summer after fruit harvest to decrease canker-causing diseases).

Growth responses to pruning tend to be strongest when trees are pruned in the winter dormant season because winter pruning removes many potential growing points (vegetative and reproductive buds) without a similar reduction in storage carbohydrates and nutrients (since buds are on thin stems and the bulk of storage is in thicker structural branches and roots). Thus, dormant pruning tips the balance between potential growing points and stored resources in favor of the growing points that remain after pruning. In addition, pruning (especially heading cuts) typically disrupts the natural apical dominance mechanisms that tend to control shoot growth. Also, heavy pruning into older branches stimulates growth of epicormic shoots.

If pruning is done during the growing season (summer pruning), it generally results in less stimulation of regrowth because trees expend much of their stored carbohydrates to generate new growth in the spring. When leafy shoots are removed, those shoots cannot provide a return on investment and thus the overall carbon/nutrient balance of the tree is reduced. The closer the summer pruning is done to the dormant period in the late summer or fall, the more likely the pruning response will be similar to dormant pruning.

Understanding the Root Sink 8

Root Growth

Root development and growth is similar to shoot growth in that extension growth is initiated by an apical meristem and girth growth of mature roots is carried out by the vascular cambium. However, the initiation of lateral roots is entirely different than the initiation of lateral leaves or shoot meristems.

The apical meristem of new growing roots is covered by layers of cells forming the root cap (Fig. 8.1) and these cells secrete a gelatinous matrix that helps lubricate the extension of the root through the soil as the root grows. The outer cell layers of roots make up an epidermis that acts as a protective covering. Inside the epidermis is a layer of tissue called the root cortex. Water and dissolved nutrients can flow outside of cortex cells and the root cortex cells function to absorb water and nutrients that can be transported from cell to cell toward the center of the root. Two inner rings of cells on the inner border of the cortex are the pericycle and the endodermis. The pericycle is a sheath of cells inside the root cortex that surrounds the center core that will eventually develop into the conducting tissues of young roots. The endodermis is a continuous ring of cells outside the pericycle that provides a functional separation between the cortex and the stele of the root.

The endodermis functions as a functional gateway between the cortex and the stele. The transverse walls of endodermal cells are suberized to prevent extracellular passage of water and dissolved nutrients from passively moving from the cortex layers to the xylem and phloem inside the stele. Thus, all water and dissolved nutrients must be taken up inside cortex cells and move intracellularly from cell to cell through the cells of the cortex and endodermis to enter the stele. The stele is made up of cells that develop into functional xylem and phloem that serve the same purposes as xylem and phloem in the aboveground portions of the tree. Mature woody roots have structures very similar to stems with a vascular cambium and xylem and phloem and an outer covering that comprises the bark.

There are no functional equivalents to nodes or metamers in roots and the initiation of lateral roots occurs at seemingly random intervals from the pericycle

© T.M. DeJong 2022. *Concepts for Understanding Fruit Trees* (T.M. DeJong)
DOI: 10.1079/9781800620865.0008

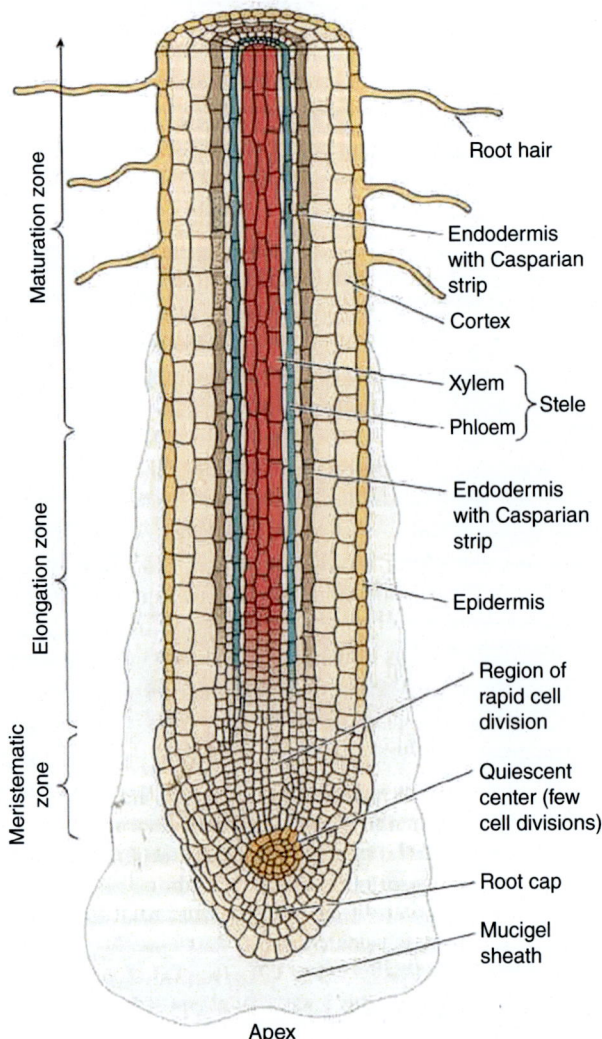

Root hair

Endodermis
with Casparian
strip

Cortex

Xylem
          } Stele
Phloem

Endodermis
with Casparian
strip

Epidermis

Region of
rapid cell
division

Quiescent
center (few
cell divisions)

Root cap

Mucigel
sheath

Maturation zone

Elongation zone

Meristematic
zone

Apex

Fig. 8.1. Diagram of a longitudinal section of a root apex. The meristematic region is near the tip behind the root cap. In the elongation zone, cells produced by the meristem elongate and differentiate to produce the xylem, phloem and cortex as well as other components of the roots. The pericycle is the thin layer of cells between the brown cells labeled endodermis and green line of cells labeled phloem. Note that root hairs are simply extensions of epidermal cells. (From Taiz *et al.*, 2015.)

inside the roots as young roots elongate. Although there are no identifiable anatomical indicators for where lateral roots arise, detailed studies of the initiation of lateral roots show that they are not initiated at random but there are complex signaling mechanisms involving external and internal stimuli that trigger lateral root development.

Roots are often classified into two broad categories, skeletal roots and fibrous roots, according to size/function. Skeletal roots are primary roots but can include higher-order roots and they form the major framework of the root system analogous to the scaffolds in the aboveground part of the tree. Skeletal roots are also often categorized by their orientation: horizontal or vertical (sinkers). The majority of skeletal roots in irrigated orchards in California are "horizontal" and rarely are found deeper than 1 to 1.5 m below the soil surface (Fig. 8.2). However, some skeletal roots appear to be "programmed" to grow vertically into the soil and can extend to further depths. If orchards are on deep soils and unirrigated, many fruit trees can grow functional roots to depths of

Fig. 8.2. Photo of an intact apple tree on display at the East Malling Research Station (UK) showing distribution of major roots. Note that most major roots would not have extended more than 1.0 m into the soil and the major structural roots are distributed horizontally. (Photo by T.M. DeJong.)

greater than 3 m. Although it is a common notion that tree roots don't extend much past the drip line of a tree's canopy, this is not true. Tree roots can extend substantial distances from the trunk of a tree if soil conditions, water availability and nutrients are present to stimulate root activity.

Fibrous roots can be further categorized by function and include axial and absorbing (or feeder) roots. Absorbing roots are fine roots whose main function is to absorb water and nutrients. The primary function of axial roots is to transport absorbed water and nutrients to the main parts of the plant. Some of both these functions can be carried out by the same roots so it is often not possible to visually distinguish differences between them. They are often initially white and extend rapidly through the soil. Absorbing roots are characterized by their high physiological activity and account for the highest rootlet numbers during active tree growth. Collectively these roots on a single tree can be miles long. Some of these roots slough off their white cortex cells, turn darker in color and develop into conducting roots, and may develop secondary thickening and eventually become skeletal roots.

As mentioned above, one of the primary functions of roots is to absorb water and minerals from the soil to support plant growth. Most of these functions are carried out in recently developed, fine roots. Fine absorbing roots can grow very rapidly, be found in soil in high concentrations, especially in regions of the soil where nutrients are plentiful, and be relatively short-lived. Many of these roots can have lifespans of less than a month while others develop a protective layer of suberin and can remain functional for longer periods of time.

A second major root function is anchorage. The skeletal structure of larger roots is very important for tree anchorage. With expansion of orchards into various soil types and windy environments, this role of roots has been receiving greater horticultural attention. Some new rootstocks are being promoted for Californian nut crops specifically for their improved anchorage. Skeletal tree root structure can vary greatly with rootstock. Some rootstocks tend to grow relatively sparce, thick skeletal roots while others can have numerous relatively thin skeletal roots. Growers sometimes talk about taproots (roots that extend down directly in line with the central axis of a tree) but commercial fruit trees rarely have true taproots because many trees are not developed from seedlings and those that are are generally "undercut" when they are removed in a nursery, so any deep roots are pruned off in the process.

A third major function of roots is carbohydrate and nutrient storage. This occurs primarily in the xylem and phloem parenchyma cells of the larger structural roots. The concentration of non-structural carbohydrates in the woody tissues of tree roots tends to be 10 to 20% higher than in comparable tissues in scions. The reason for this is not well understood but one possibility is that, because roots are imbedded in soil, they do not need as much of their anatomy devoted to providing structural strength (such as fibers) as the aerial portions of the tree. Thus, the ratio of storage parenchyma cells to other cell types in roots may be higher than in aerial portions of the tree. Roots should be considered a major long-term storage organ of trees.

Root growth is sensitive to temperature and where soils are cold, root growth is very slow. Roots can also be sensitive to high temperatures so if trees are growing in hot climates where there is a lot of direct sunlight, root growth will be inhibited near the surface of exposed soil. Unlike the upper part of trees, roots are not thought to experience endodormancy (true dormancy) and thus don't have to experience long periods of chill before being stimulated to grow in the spring. Thus, it is possible to stimulate root growth in the middle of the winter on a "dormant" tree by simply heating up the soil. However, this growth will be limited because there is no need for active root water and nutrient uptake if the top of the tree is not active.

The seasonal timing of root growth varies somewhat with species and probably depending on the conditions under which it is measured. It is widely accepted that there is generally a strong flush of new root growth in the spring as soils warm up and the tops of trees are also growing rapidly (Fig. 8.3). This is logical since new shoot growth and leaf activity require major increases in root water and nutrient uptake. Root growth during the summer period can be more variable. If trees have heavy crop loads, root growth rates generally decrease due to competition for photosynthates by the fruit and, because shoot growth rates also tend to decrease with heavy crop loads, there is decreasing demand for water and nutrients (Fig. 8.4). Root growth has been reported to show a small flush after harvest of crops with some species, but the extent of this flush is likely dependent on the relative change in tree nutrient and water status before and after harvest as well as environmental conditions.

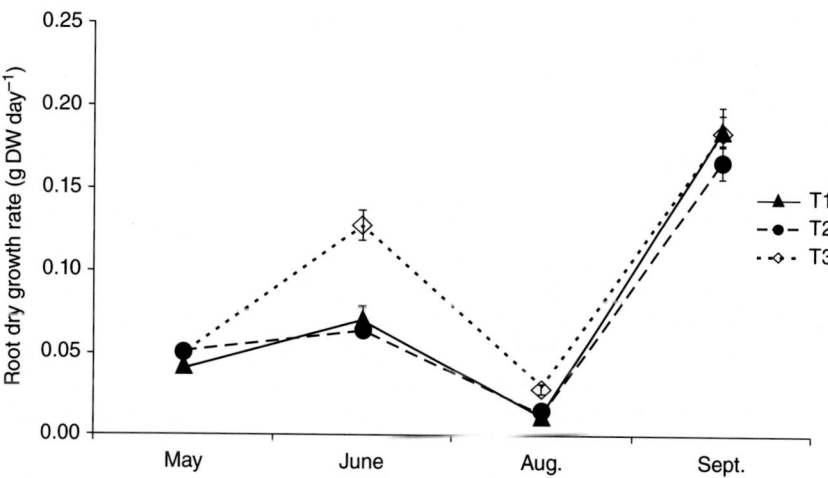

Fig. 8.3. Patterns of peach root growth rate (DW, dry weight) measured in ingrowth root bags during four periods of a growing season. Trees were subjected to three different cropping treatments (T1, unthinned; T2, commercial crop load; T3, defruited). Trees were harvested in August. Symbols indicate mean values and vertical bars indicate standard errors of the means. (From Ben Mimoun and DeJong, 2006.)

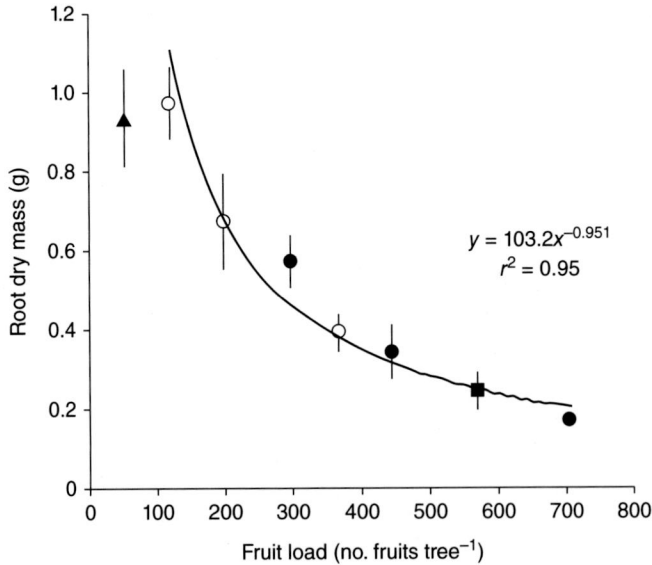

Fig. 8.4. Comparative increases in root mass in ingrowth root bags on peach trees subjected to various thinning treatments that resulted in different crop loads. Increases in root mass were measured over the Stage III period of fruit growth. Note the strong reductions in root growth as crop loads increased. Symbols indicate mean values and vertical bars indicate standard errors of the means. (From Marsal *et al.,* 2003.)

Root growth can be negatively affected by lack of water in the soil, but growth and activity can be also very sensitive to the concentration of O_2 in the soil. Heavy soils (with high clay content) tend to be easily waterlogged and in waterlogged soils pore spaces between soil particles are filled with water instead of air. Under these conditions roots can be starved for O_2 and die. Waterlogged soils are particularly problematic if soils are warm and the tree is actively growing or beginning to grow. Waterlogged soils are generally not a problem in the wintertime because the tops of the trees are dormant and soils are colder. Even in California, which has a relatively dry climate, many orchards suffer more from mismanagement of irrigation causing lack of soil aeration and consequent root stress than lack of water. The susceptibility of roots to anoxia varies greatly among fruit tree species and rootstocks. Some pear rootstocks are very tolerant to waterlogged soils, while almond, which is native to dry environments, is very sensitive to them.

Soil compaction also has large negative effects on root growth and activity. Compact soils can make it difficult for roots to penetrate areas of the soil, as well as decreasing pore spaces between soil particles so the soil is less aerated. Soil compaction is made worse by repetitive passes through orchards with heavy equipment. Soil compaction can be mitigated by adding organic matter to the soil as well as planting cover crops.

Pruning of the upper part of the tree can tend to decrease root growth and vice versa. Root pruning can be used to decrease the vigor of the upper part of the tree. These responses are logical when we consider that heavy pruning of the tops of trees tends to stimulate increased vegetative regrowth in the top of trees and this regrowth competes for photosynthates going to the roots, and when the tops of trees are temporarily smaller, they need fewer resources from the roots. Similarly, root pruning stimulates rapid root regrowth that competes for resources required for growth of the top and reduces supply of resources coming from the roots because root uptake is temporarily limited until the roots regrow. Root pruning is only practiced in special circumstances where decreased fruit size is not a problem because root pruning can substantially reduce fruit growth.

Rootstocks

Almost all commercially available fruit trees are compound trees with scions (top part) that are genetically distinct from the roots. This allows for clonal propagation of scion cultivars by grafting or budding scions on to rootstocks. It also provides for rootstocks to be selected for desirable rootstock characteristics such as tree vigor, anchorage, graft compatibility with various scion cultivars, resistance/tolerance to soil-associated diseases, performance in specific soil types, nutrient uptake, resistance/tolerance to drought or water-logged soils, etc. When purchasing trees for planting an orchard, the grower should pay attention to differences in rootstock characteristics as well as scion characteristics.

In recent years, there has been increased interest in rootstocks that can reduce the height of the scion so that trees can be managed with a minimum of ladder work and hand labor. This has been most successful with apple production and the apple industry worldwide has been revolutionized by the introduction of size-controlling (dwarfing) rootstocks. Size-controlling rootstocks are now available for peaches, plums, cherries and other fruit crops. The physiological mechanism underlying vigor control in the scion by specific rootstocks has been a subject of research for nearly 100 years. Early research indicated that it may have something to do with the xylem anatomy of specific rootstocks but later this notion was discarded after researchers began studying internal plant signaling compounds such as plant hormones. While there is ample evidence that concentrations of plant hormones differ in association with growth characteristics of scions on dwarfing versus vigorous rootstocks, there is little evidence to clarify whether these differences are a cause or an effect of the differences in vigor. Other research has implicated that the graft unions in specific scion/rootstock combinations may influence the movement of nutrients or carbohydrates up the tree from the rootstock.

A mechanism for explaining the rootstock-mediated size-controlling response that has recently received a lot of support has circled back to original

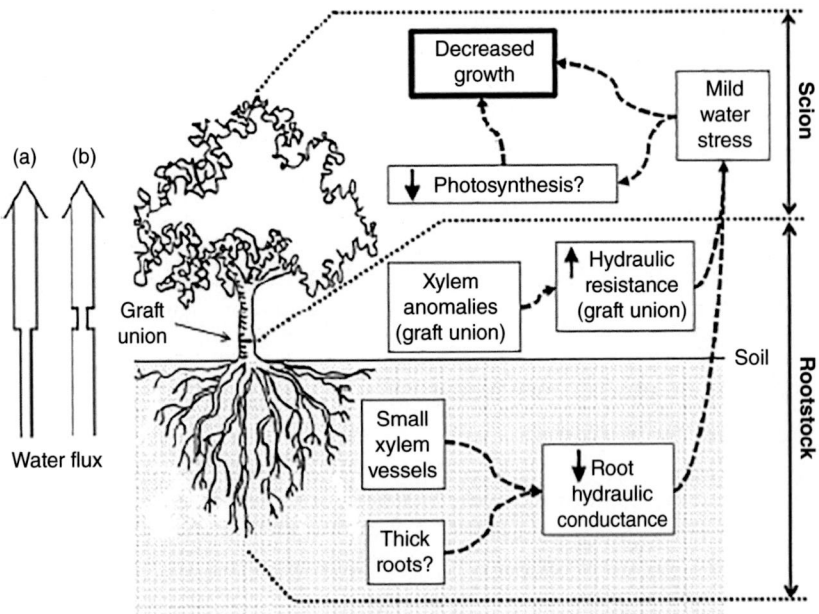

Fig. 8.5. Schematic illustration of the water relations-mediated size-controlling mechanism for trees grafted on size-controlling rootstocks. The arrows on the left side indicate that the restrictions to water movement may be a function of limitations to hydraulic conductance in the whole root system (a) or just the graft union (b). (From Basile and DeJong, 2019.)

reports that the xylem anatomy of the rootstock is likely the factor involved in eliciting the size-controlling behavior of specific rootstock/scion combinations. It has been shown that select size-controlling peach rootstocks have average xylem vessel diameters that are smaller than more vigorous rootstocks. It is thought that, in the springtime when new xylem vessels are being developed and the bulk of water conduction in the rapidly growing trees is dependent on recently formed xylem, the smaller average vessel diameter of the new vessels in dwarfing rootstocks restricts the conduction of water up the trees. This restriction is enough to cause small reductions in the water status and growth rate of growing shoots (Fig. 8.5). When this effect is compounded over several weeks during the "grand period of growth" in spring, it results in less vigorous trees and especially less growth of epicormic shoots. Once reduced growth is realized early in tree development, it results in less need for heavy pruning and this, in turn, results in less stimulation of epicormic shoots, since epicormic shoots are produced in response to pruning. Thus, the size-controlling effect is compounded over time by differences in xylem anatomy and orchard management practices.

Understanding the Fruit Sink 9

Floral Development

All fruits and nuts are derived from the outgrowth of specific parts of flowers or inflorescences (structures that support numerous flowers). In almost all temperate deciduous fruit trees, floral initiation and differentiation occur in the season prior to flowering. Floral initiation occurs when a "floral signal" is received by a terminal or lateral meristem that will eventually develop into a flower or inflorescence bud. This occurs as early as May of the year preceding bloom in some species. Although floral induction or initiation occurs when a "floral signal" is received by a meristem, the event is not visually detectable until later when floral differentiation begins. However, we can assume that floral induction occurs weeks or even months before visual signs of floral differentiation because, in some cases, the presence of immature fruit on a shoot in late spring can inhibit the appearance of flower buds on that shoot later in the summer. Whereas removal of the same fruit in early spring results in the appearance of flower buds later that summer.

Floral differentiation begins when parts of the flower or inflorescence can begin to be microscopically distinguished in a floral meristem (Fig. 9.1). In most temperate deciduous fruit tree species, this occurs from July to September in the year prior to bloom (Fig. 9.2). By late fall, all of the flowers that will bloom the following spring are well developed before the tree goes into endodormancy. Most people believe that virtually no growth occurs during the dormant season but in some species very slow development and growth of the floral organs can occur throughout the winter months and more rapid growth can precede bloom (Fig. 9.3).

Floral anthesis (bloom) occurs in the late winter or early spring after dormancy is broken and the flower or inflorescence resumes more active development and growth. Among the various tree species there is a wide array of types of flowers or inflorescence structures that can develop into fruits, nuts or products of commercial significance. Some fruits consist only of products of floral ovary development (stone fruits and true berries) while others consist

© T.M. DeJong 2022. *Concepts for Understanding Fruit Trees* (T.M. DeJong)
DOI: 10.1079/9781800620865.0009

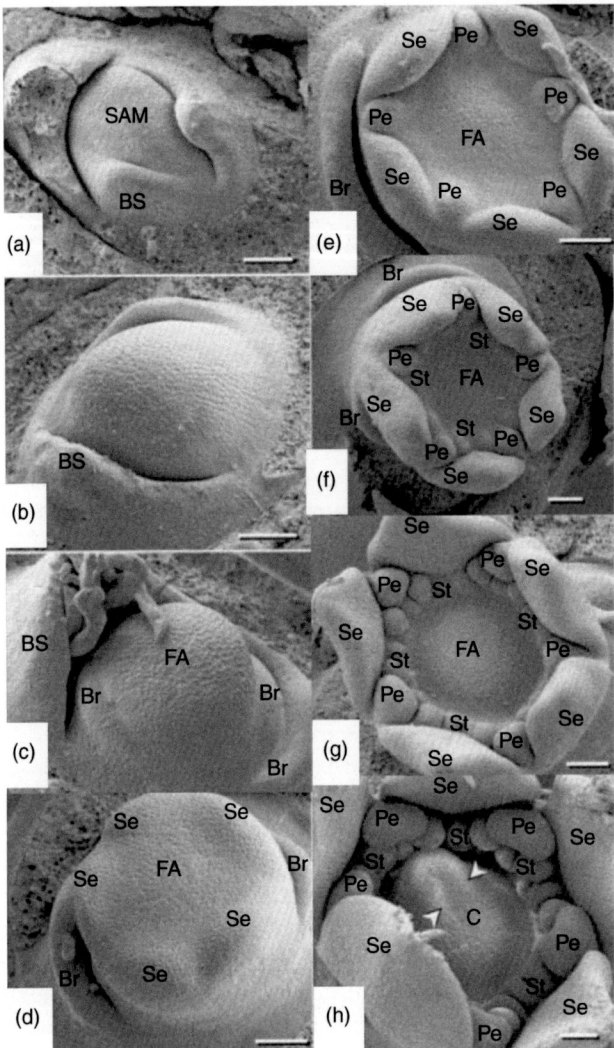

Fig. 9.1. Scanning electron micrographs showing the progressive development of almond floral buds from beginnings as a shoot apical meristem (SAM) with bud scales (BS) (a), to the transition to a floral meristem (b), to a floral apex (FA) with three bracts (Br) (c), to the initiation of sepals (Se) (d) and petals (Pe) (e) and stamens (St) (f), and finally the carpel (C) (h, note arrowheads). (From Lamp *et al.*, 2001.)

of both the floral ovary and the structures that it is imbedded in (pome fruits). Some reproductive buds only contain a solitary flower (peach, almond) while others develop into inflorescences bearing multiple flowers (apple, pear, pistachio) and in some cases the main edible part of the fruit consists of the inflorescence (strawberry, figs).

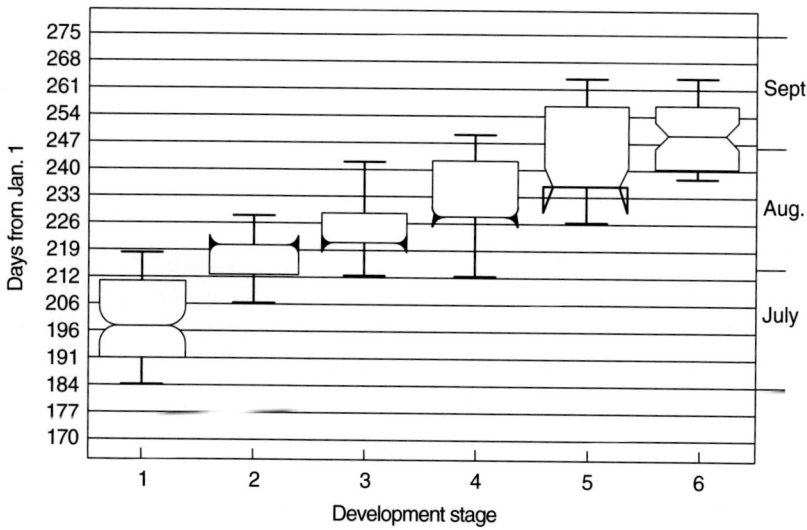

Fig. 9.2. Box-and-whisker plots showing the progression to different stages of 'Non-pareil' almond bud development in Davis, California, in 1997. Note that floral bud development occurs over two months from July to September. Harvest of nuts from previous-year floral development occurs in the same time period. Boxes indicate 95% confidence intervals for the median values. (From Lamp *et al.*, 2001.)

Pollination, Fertilization and Fruit Set

Most fruit and nut species require pollination (transfer of pollen from floral stamens (male structures) to stigmas (female structures)) and fertilization (fusing of male sperm cells with female egg cells) in order to initiate fruit set, development and subsequent growth. However, specific varieties of a few species can achieve parthenocarpic fruit set without pollination and fertilization under specific conditions (e.g. Bartlett pear in specific climatic regions and Mission figs).

Fruit and nut trees are biological systems and have been selected to be able to adapt to different or changing environments. Nature has tended to favor systems that foster variability in their offspring. One means of achieving this is to have pollination systems that foster cross-fertilization among individuals in populations of plants. Variability in genetic make-up of individuals and populations of plants is favored so that changing plant populations can adapt to changing or new habitats. Thus, many fruit species have pollination systems that favor out-crossing among individual plants. Some fruit tree species have self-incompatible pollen so that pollen from the same individual cannot pollinate flowers of the same individual (cherries, almonds, specific plums, etc.). Some apple varieties are mainly self-incompatible but become partially self-compatible as flowers age. Other species have separate male and female flowers on the same plant (dichogamy; walnut, pecan) and the timing of pollen

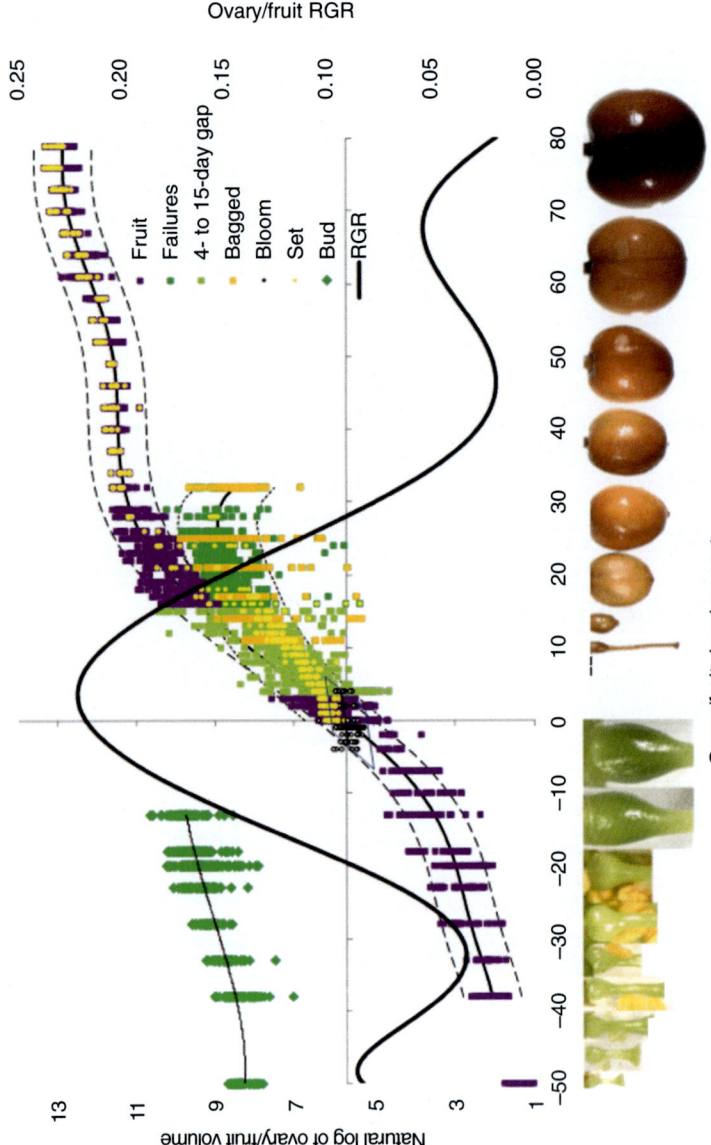

Fig. 9.3. The pattern of sweet cherry pistil/fruit accumulation in volume and relative growth rate (RGR) from 50 days before bloom to fruit maturity. Note that a substantial amount of growth occurs before bloom and the fruit relative growth peaks near bloom and declines rapidly thereafter. (From T. Einhorn, D.M. Gibeaut and M. Whiting, Oregon State and Michigan State Universities, 2020, personal communication.)

shedding occurs before (protandrous) or after female flower receptivity (protogynous) (Fig. 9.4). Still other species have separate male and female plants such as pistachio and kiwifruit. Even if trees are pollen self-compatible most require insects or wind to move pollen from the stamens to the stigmas of the same or different flowers on the same plant (select plums, apricots, cherries and almonds). On the other hand, most peach varieties have flowers that

(a)

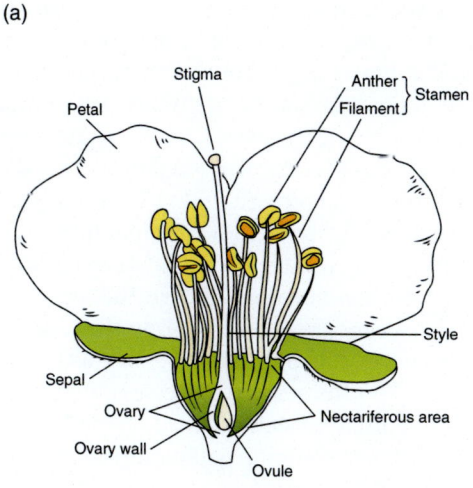

(b)

Fig. 9.4. The structure of a (perfect) stone fruit flower with both male and female flower parts (a) and dichogamous flowers of walnut with separate female and male flowers (b). (a) Diagram of a peach flower. (From Polito *et al.*, 2012.) (b) Walnut female (pistillate) flowers borne on mixed terminal buds and male (staminate) flowers borne on catkins. (From Polito, 1998.)

are self-pollinating (cleistogymous) and pollen gets transferred as the stigma extends up through the anthers as the flowers open.

Floral pollination, fertilization and initial fruit set are among the most critical, variable and unpredictable processes that influence successful fruit and nut production. To be successful, pollen development and maturity must coincide with periods of floral stigma receptivity. After pollen transfer to a stigma, it germinates, and the pollen tube must grow down the floral style to an ovule in the floral ovary while the ovule is still receptive. Each of these processes takes time and their rates of development and growth are dependent on environmental conditions, particularly temperature. Furthermore, the temperature responses may vary among the various processes. What contributes to make pollination, fertilization and fruit set so variable is the fact that early spring weather conditions and temperatures tend to be the most variable of any time during a given year. If these processes are not successful or are only marginally successful, there is nothing that can be done to increase fruit yield in that year and, practically, much of the year's solar energy collection is for naught if fruit set is poor. The narrow window during which pollen transfer, germination, growth to the ovule and fertilization occur has been described as the "effective pollination period" (Fig. 9.5). If pollen is not transferred during this critical time, pollination and fruit set are not successful, and the length of this period can vary with temperature, which can affect the various processes differently.

Many temperate deciduous fruit and nut tree species are insect pollinated (including pome and stone fruit) and the most common pollinating insects are bees. This can further complicate and add to the unpredictability of fruit set because weather can also strongly inhibit bee activity, especially cold temperatures and rain. Other species are wind pollinated and again the presence of wind is weather dependent.

Flower Structures and Fruit Types

Flowers can be borne singly from one floral bud as in peach and almond, in groups of two or three from one floral bud as with plums and cherries, or in small inflorescences from compound reproductive buds that have preformed vegetative and reproductive parts that develop into a short shoot with a single (quince) or multiple flowers (apples and pears). Some species such as pistachios and grapes develop flowers from compound floral buds in large inflorescences with numerous individual flowers (Fig. 9.6). Individual flowers are categorized by their floral anatomy and where the ovary is positioned relative to the attachment of the other floral parts. With hypogynous flowers the floral sepals, petals and stamens are attached to the floral receptacle below the pistil (gynoecium) and the fruit grows to form true berries (persimmon, kiwifruit) (Fig 9.7a). With epigynous flowers the floral sepals, petals and stamens are attached to the floral receptacle above the gynoecium and the wall of the gynoecium is fused with the receptacle to form pome fruit (apple, pear, quince) (Fig 9.7b).

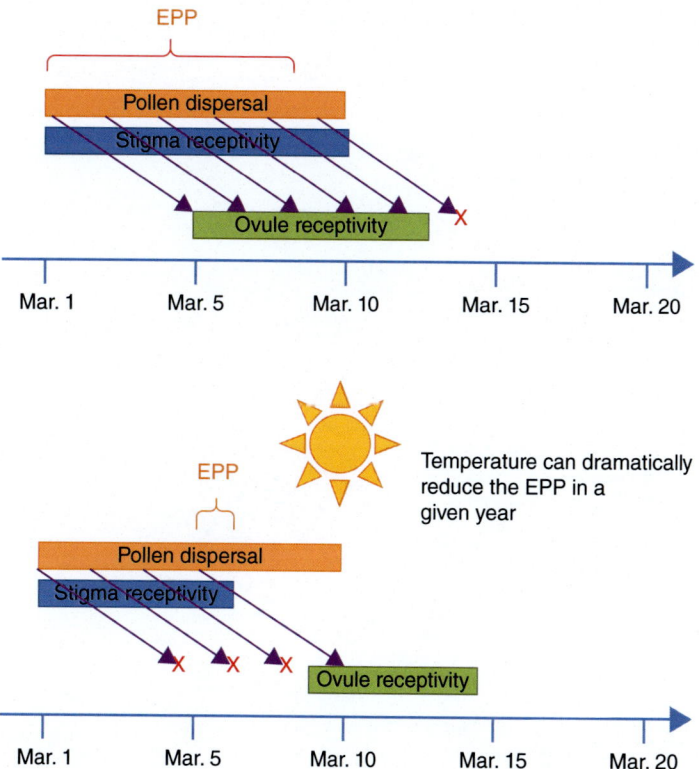

Fig. 9.5. The effective pollination period (EPP) is critical for successful pollination and fertilization. EPP can vary depending on temperature. Pollen dispersal, stigma receptivity, ovule receptivity and pollen tube growth can all be affected differently by temperature. This diagram illustrates differences in the EPP under good conditions and if conditions shorten stigma receptivity and delay ovule receptivity. (From V.S. Polito, UC Davis, 2014, personal communication.)

Perigynous flowers have an intermediate type of structure in which the sepals, petals and stamen tissues originate below the gynoecium but remain fused to form a floral tube, so the floral tube surrounds the gynoecium until it extends above the floral tube creating a stone fruit (peach, cherry, almond, walnut, etc.) (Fig 9.7c). The floral tube of perigynous flowers is not fused with the gynoecium and in many species it has glands that secrete nectar to attract insects for pollination.

Stone fruits or drupes (peaches, plums, apricots, cherries, almonds, walnuts, pecans, hazelnuts) are derived from perigynous flowers with a simple pistil that has a single carpel (ovary). The edible, fleshy part of stone fruit is the mesocarp, which is derived from the carpel wall. The epidermis of the fruit is termed the pericarp. The stony, lignified part of the fruit (pit) is derived from the inner wall of the carpel and is designed to protect the seed that

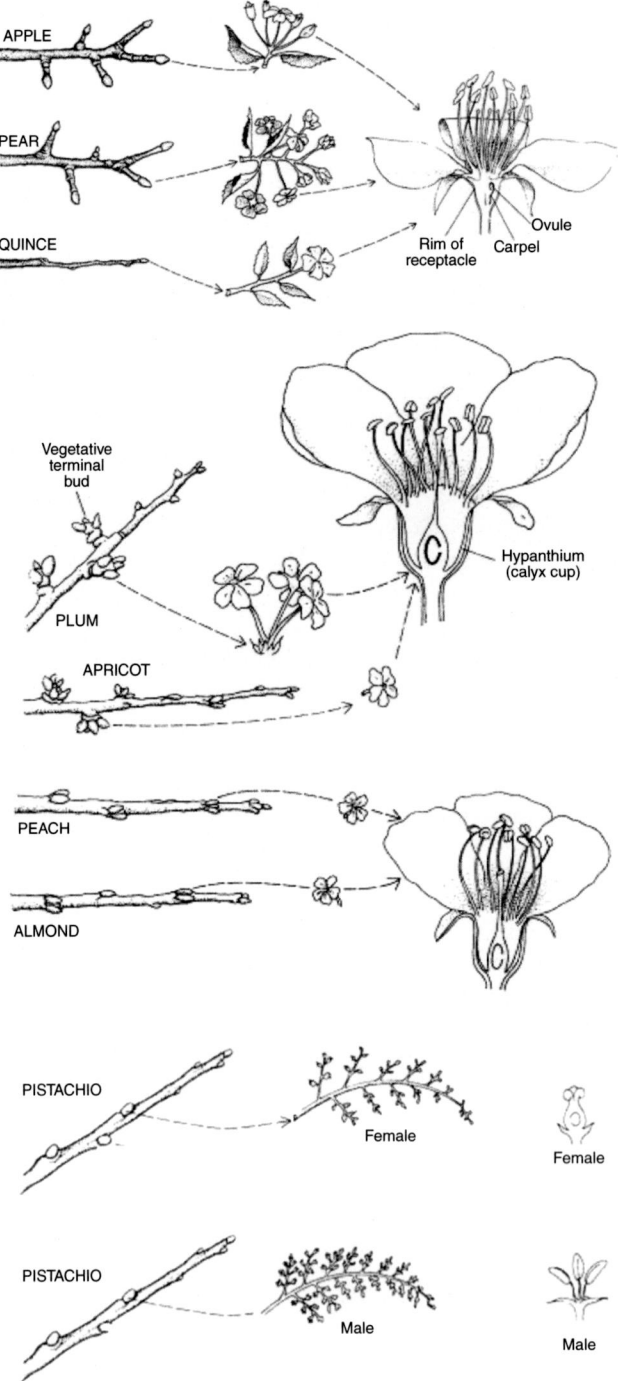

Fig. 9.6. Bearing habits of pome fruits (apple, pear, quince), stone fruits (plum, apricot, peach, almond) and pistachio. (From Westwood, 1978.)

(a)

(b)

(c)

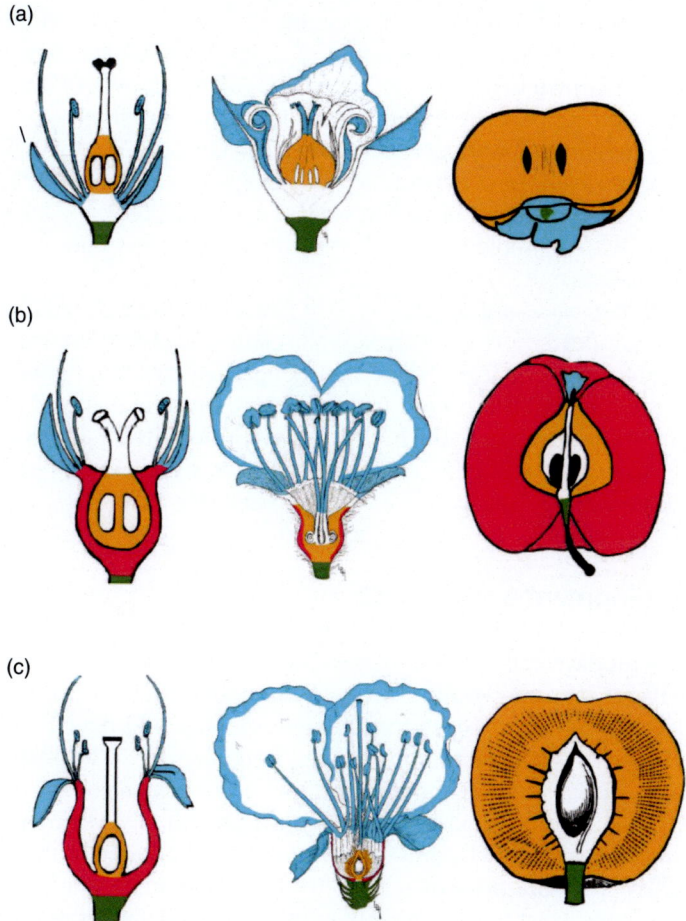

Fig. 9.7. Different flower types develop into different types of fruit according to location of attachment of the other floral organs (perianth and stamens) to the receptacle relative to the gynoecium: hypogynous flowers develop into true berries (a); epigynous flowers develop into pome fruit (b); and perigynous flowers develop into stone fruit (c). (From UC Davis Fruit and Nut Research and Information Center, 2021.)

develops as a result of the fertilization of the egg within the ovary. Ecologically, the purpose of the mesocarp of stone fruits is thought to be to attract animals to eat the fruit and carry them away from the parent tree, thus dispersing seeds. Olives are a special type of drupe derived from hypogynous flowers.

Pome fruit (apple, pear, quince) are derived from epigynous flowers that generally have compound pistils with five carpels (ovaries). Each carpel can have two ovules and thus, when fully fertilized, the fruit will have ten seeds. Most of the edible tissue of a pome fruit is derived from the floral receptacle and the outline of the core of a pome fruit demarks the location of the ovary walls.

Differences in the texture of the edible portion of stone and pome fruit can be attributed to the differences in the origin of tissues from which they are derived. Seeds are known to promote the development and growth of pome fruit and thus the number of seeds in the fruit can influence fruit shape.

Most hypogynous flowers produce fruit called berries. The edible flesh of berries is derived from the ovary wall of the hypogynous flowers and contains no lignified tissues other than those in seeds. These fruits typically have compound pistils with multiple carpels. Persimmons, kiwifruits and grapes are examples of berries.

Figs and mulberries are a special type of fruit called multiple fruit because they represent inflorescences and are actually composed of multiple individual flowers fused to form one fruit. In the case of fig, the individual flowers/fruitlets are borne "inside" the overall fruit structure and the external portion of the fruit is the structure to which the flowers are attached (thus, it is a kind of "inside out" multiple fruit). Even more odd is the fact that each fruit can have separate male and female flowers within it (Fig. 9.8).

Fruit Development and Growth

Aspects of fruit development and growth have been studied extensively over the past several centuries. Fruit growth of many tree fruit species has been described as falling into two patterns, single sigmoid and double sigmoid, when fruit growth is studied as accumulated fruit diameter or by fresh or dry weight. Fruit with a single-sigmoid growth pattern (such as pome fruits and most nuts) start with a period of relatively slow growth followed by a period of more rapid, steady growth and then a final period of slow growth when the fruit approaches maturity (Fig. 9.9). Fruit with a double-sigmoid pattern (many stone fruits) start with a period of relatively slow growth that gradually increases (Stage I) and then growth tends to slow down for a period (Stage II) before it resumes rapid growth (Stage III) and finally levels off again (Stage IV) as the fruit approaches maturity (Fig. 9.9). The last stage of growth (Stage IV) is often not observed because many stone fruits are harvested prior to reaching this stage. Plotting fruit growth as an absolute growth rate (increase in fruit weight per unit time) tends to accentuate differences in periods of rapid growth from periods of slower growth and shows the first three stages of peach fruit

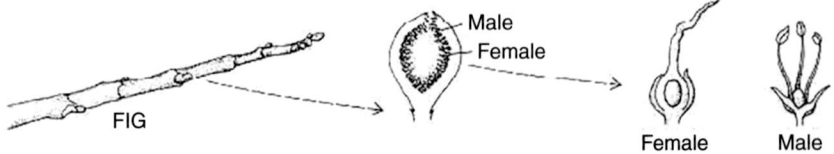

Fig. 9.8. Diagram of a fig fruit, which is an "inside-out" inflorescence with female and male flowers inside the fruit. (From Westwood, 1978.)

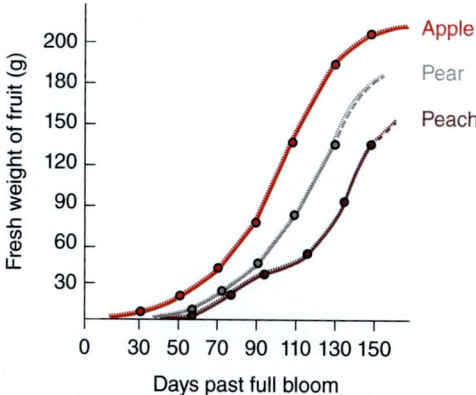

Fig. 9.9. Apple and pear have a typical sigmoid growth curve except that the last part of the pear curve is not apparent because the fruit is usually harvested before the final plateau is reached. Late-maturing peaches have a double-sigmoid growth curve, but early-maturing peaches have a curve very much like pear because the period of fruit growth is shorter and there is no mid-season slowdown of growth. (From Westwood, 1978.)

growth more clearly than cumulative growth curves (Fig. 9.10). It should be noted that extra-early-maturing varieties of fruit species that normally exhibit a double-sigmoid growth curve (peaches and nectarines) do not exhibit Stage II of fruit growth. The main difference in peach fruit growth patterns of varieties that mature at different times of the growing season is the length of Stage II of fruit growth.

Fruit growth can also be described as a relative growth rate (RGR) (weight added to a fruit over a specific time interval per unit of fruit weight at the beginning of the time interval, weight/weight/time); conceptually, this is the same as a compound interest rate in banking. Mathematically,

$$RGR = \frac{1}{w}\frac{dw}{dt} = \frac{\ln w_2 - \ln w_1}{t_2 - t_1}.$$

The growth of most fruits can be described by an initially high RGR that rapidly declines and then levels off or slowly increases at a low RGR prior to approaching fruit maturity (Fig. 9.11). When fruit growth is expressed as an RGR the only real difference between fruit that have single- and double-sigmoid cumulative growth patterns is the initial rate of decline of the RGR curve and the timing of when the RGR stops its rapid descent and begins to level off.

An advantage to understanding fruit growth as a function of RGR is that RGRs are directly related to fruit respiration rates and thus are very helpful in describing the energetics of fruit growth. In addition, relative growth analysis can be used to help explain functional limitations to fruit growth and how fruit growth responds to horticultural practices such as fruit thinning, since growth in fruit weight for any given time interval is always a function of growth that

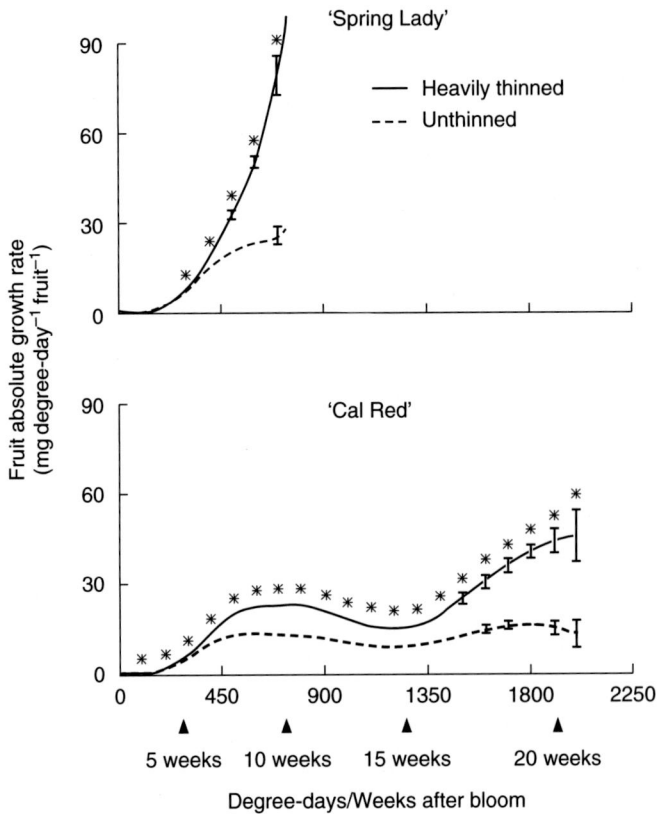

Fig. 9.10. Patterns of absolute peach dry weight fruit growth of an extra-early ('Spring Lady') and a late ('Cal Red') peach cultivar on trees that were heavily thinned shortly after bloom so the fruit could grow without resource limitations and on trees with fruit that were not thinned so fruit growth was resource limited. The three stages of 'Cal Red' fruit growth rates are very clear but the same stages are not seen in 'Spring Lady'. Vertical bars indicate standard errors and asterisks indicate when there were statistically significant differences between mean values. (From Grossman and DeJong, 1995a.)

occurred during previous growth intervals and the growth conditions of the current interval.

Interestingly, there has been much scientific discussion about the causes of the double-sigmoid fruit growth pattern of peaches compared to the single-sigmoid growth curve of pome fruit. However, a simple comparison of the fruit RGRs of both types of fruits indicates the differences in the two types of cumulative and absolute growth curves are simply a mathematical result of differences in the rates of descent in RGR curves of the two different types of fruit (Fig. 9.12). The very rapid decrease in RGR right after bloom in stone fruit results in a slowdown in cumulative growth in the middle part of the growth curve,

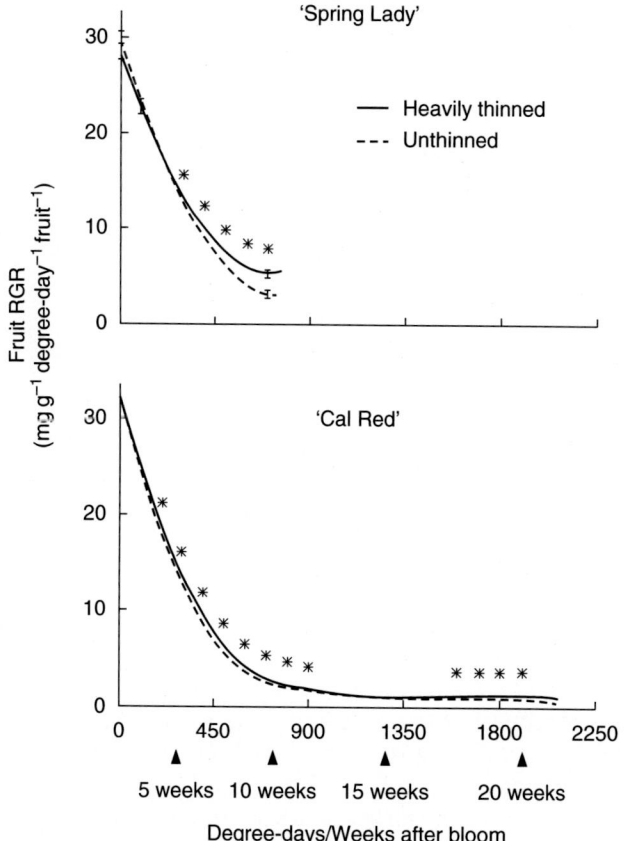

Fig. 9.11. When peach fruit growth is expressed in terms of relative growth rate (RGR) (akin to compound interest rate), the RGR is very high right after bloom and descends rapidly as the fruit grows. The solid lines in these graphs indicate the growth potential per unit fruit weight at specific times after bloom. Dashed lines indicate RGR patterns on unthinned trees. Note that the main difference between the early-maturing variety ('Spring Lady') and the late-maturing variety ('Cal Red') is the length of the fruit growth. The asterisks indicate significant differences between thinning treatments and the periods when fruit growth on trees with heavy crops were resource limited. (From Grossman and DeJong, 1995a.)

while the slower rate of decrease in RGR of pome fruit results in a continuously increasing cumulative growth curve.

Limitations to Fruit Growth

Logically, individual growth of a fruit is limited both by the inherent potential of the fruit to grow (sink limitation) and/or the availability of assimilates

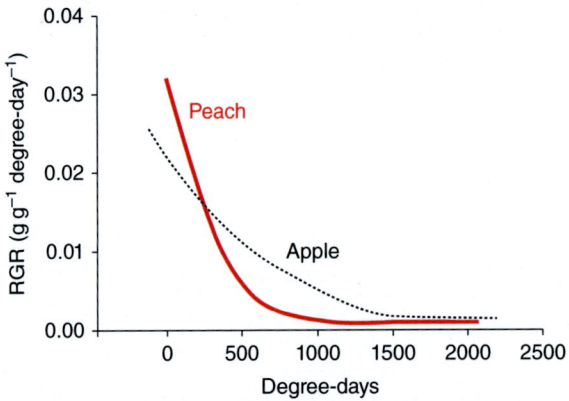

Fig. 9.12. The primary difference between peach and apple relative fruit growth rates is simply the rate at which the relative growth rate (RGR) declines after bloom. The differences in RGR patterns account for the single- and double-sigmoid cumulative fruit growth curves of apple and peach fruit, respectively. (Adapted from Grossman and DeJong, 1995; Pavel and DeJong, 1995a.)

(source limitation) during specific intervals of growth. It is not possible to ascertain which is most limiting in normally cropped fruit trees by comparing fruit mass accumulation curves or fruit absolute growth rates. However, if fruit loads on trees are reduced to very low levels very early in the growing season, it is reasonable to assume that the remaining fruit on the trees are primarily sink-limited during the entire fruit development period. Then fruit RGRs from trees with very low crop loads can be compared with fruit RGRs from trees with normal crop loads to indicate primary periods when fruit growth is limited by source limitations in normally cropped trees. Studying peach fruit growth in this manner indicates that, under commercial crop loads, source limitations often occur in the early and late stages of fruit growth while growth during the middle period is largely sink-limited. The primary effect of crop load on individual fruit size is a function of the amount of time fruit growth is source-limited compared to sink-limited.

Source limitations to individual fruit growth can be a function of either lack of total resources available in the tree to support fruit growth or transport/competition limitations imposed by transport resistances or competition from other sinks. Transport/competition limitations to fruit growth are prominent during early phases of fruit growth but overall resource supply is more important during final phases of fruit growth. This is logical since the tree must develop new phloem in the spring to transport assimilates from leaves to fruits and other sinks, and the rate of new phloem development may limit transport rates between sources and sinks early in the season.

RGR analysis has been used to show that distribution of fruits within a tree canopy can also have substantial influence on transport/competition limitations to fruit growth. Assimilates from leaves on de-fruited scaffolds of peach trees do not significantly support growth of fruits on another scaffold. Also, growth of a given number of fruits was less when the fruits were unevenly distributed among a specific number of shoots than when the same number of fruits was distributed evenly among the same number of shoots.

Since photosynthesis provides the currency to support growth of organs, there is a tendency to try to explain many experimental results in terms of treatment effects on photosynthesis. However, fruit RGR analysis has shown that the most important factor causing decreased fruit size in N-deficient peach trees tends to be a shortening of the fruit development period rather than decreased availability of assimilates during fruit growth.

Thinning Effects on Fruit Growth, Size and Yield

It is widely recognized that, in general, fruit size at harvest is inversely related to crop load and fruit yield per tree (Fig. 9.13). However, markets often place a higher value on large fruit than small fruit, thus a grower thins fruit early in the growing season and sacrifices some crop yield to attain marketable fruit sizes. Probably the most practical application of fruit RGR analysis is in using it to understand fruit growth and crop yield responses to fruit thinning. Fruit RGRs can be used to quantify the growth potential of fruit of a given cultivar for any time interval throughout the fruit development period because dry weight accumulation is expressed per existing dry weight at the beginning of the growth interval and the elapsed time during the interval. Thus, the potential RGR for a developmental time interval (which is dictated by the genetics

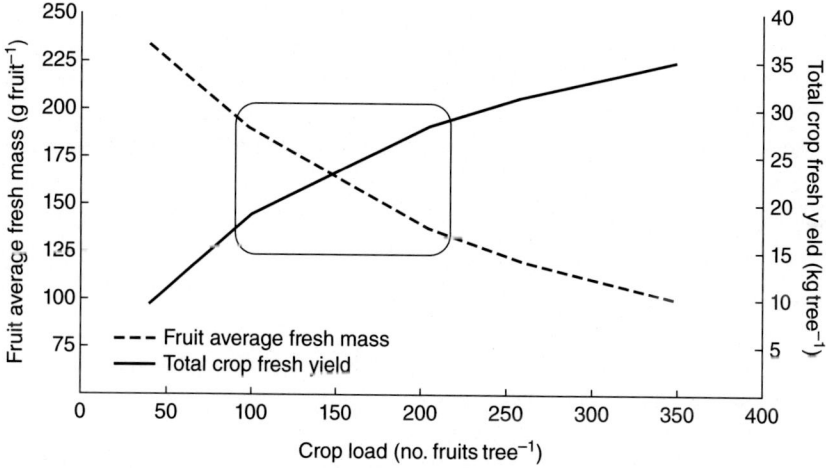

Fig. 9.13. Increasing crop loads result in greater total crop yields but smaller fruit size. The grower's job is to optimize crop loads to result in crop yields that optimize marketable sized fruit (box). (Simulated data derived from running the L-Peach model. Adapted from Lopez *et al.*, 2010.)

of the cultivar) is equivalent to the potential "interest rate" compounded on the dry weight at the beginning of an interval. Growth is compounded based on the potential RGR, the dry weight at the beginning of the interval and the ability of the tree to supply the resources to achieve the potential RGR. If the potential RGR is not achieved, the actual growth is less than the potential and both the potential and the actual growth for every subsequent interval are less than they could have been because the growth for each subsequent interval is based on the dry weight at the beginning of that interval. This means that any potential fruit growth that is not achieved early in the season cannot be made up later. Thus, final fruit size is not only a function of crop load near harvest but also of crop load throughout the season.

It is important to point out that these RGR analyses are carried out in terms of dry weight because fruit dry weight more accurately reflects the energetic inputs into fruit growth than analyses using fruit fresh weight. Water is the main constituent of fresh fruits. Water availability to fruit is not directly dependent on photosynthesis and fruit fresh to dry weight ratios can vary over a growing season. Thus, analysis of fresh fruit growth can be misleading for studying carbohydrate distribution in fruit trees. However, estimations of fruit fresh weight from fruit dry weight calculations can always be made with data on fruit fresh weight to dry weight ratios, for communicating horticulturally relevant results.

The behavior of dry weight fruit growth has been verified with whole-tree, fruit thinning experiments of early- and late-maturing peaches in California (Fig. 9.14), crop modeling studies and commercial orchard thinning trials. Fruit thinning trials in commercial California peach orchards showed that thinning earlier than was previously recommended for commercial clingstone peach orchards in California can result in either larger fruit sizes at similar crop loads or similar fruit sizes at higher crop loads (Table 9.1), without any significant increases in fruit defects such as split pits. These results with peach are consistent with thinning research in many other fruit tree species and it is safe to generalize these general principles derived from peach thinning experiments to most other fruit crops.

Spring Temperature Effects on Fruit Size

The period of fruit development (time from bloom to harvest) of a particular cultivar can vary annually by as much as a month. In some years, harvest can be as much as two weeks earlier than average while in other years it can be two weeks later than average. When the fruit development period is exceptionally short and harvest is early, usually fruit sizes are also smaller than normal. Reductions in fruit size associated with shortened fruit development periods also can be explained by RGR analyses. The length of the fruit development period for cultivars in a specific year is a function of the general genetically

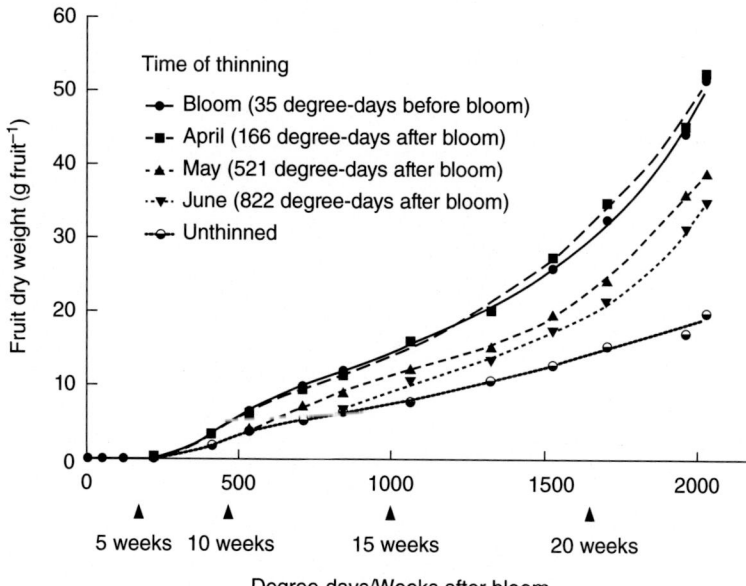

Fig. 9.14. The results of a peach fruit thinning study. Whole trees were thinned at specific times relative to bloom time. Data show the patterns of fruit dry weight as the season progressed. Note that, as predicted by relative growth rate analysis, fruit on trees thinned at later dates never achieved the mean fruit sizes of early thinned trees, even though final crop loads were the same. (From Grossman and DeJong, 1995b.)

determined pattern of growth for the cultivar and the temperatures experienced in the field during the first 30–50 days after bloom.

Temperature dependence of the length of the fruit development period has been successfully quantified by calculating the cumulative growing degree-hours during the first 30 days after full bloom (GDH_{30}) for several Californian peach, nectarine and plum cultivars over numerous growing seasons (Fig. 9.15). Furthermore, most of the temperature-dependent differences in the time between full bloom and the date of fruit maturity among years for a specific cultivar can be accounted for by differences in the time between full bloom and pit hardening in stone fruits.

Counterintuitively, in years when early spring weather is warm and temperatures are high (GDH_{30} > 6000) there is a strong tendency for fruit sizes to be small, while fruit sizes tend to be larger in years when early spring temperatures are cooler (GDH_{30} < 6000). These trends in fruit size responses among years have been explained by running fruit RGR models for springs with high and low spring temperatures. These models indicate that in springs with high early temperatures, not only is the total fruit development period of a given cultivar shortened but also the daily assimilate requirements for individual fruit growth are much higher, earlier in the season.

Table 9.1. Results of a peach fruit thinning trial documenting that early fruit thinning can increase fruit sizes and/or crop yields. The study was conducted on early-maturing ('Loadel' and 'Carson'), mid-season ('Andross') and late-season ('Ross') processing peach varieties. Whole trees were thinned on two dates. As predicted from RGR studies, early thinning resulted in higher yields with equal or greater mean fruit sizes and crop loads. Data are mean values ± standard errors. (From DeJong, 2012.)

Cultivar/Thinning date	Fruit size (g FW fruit^{-1})	Crop load (no. fruits tree^{-1})	Yield (ton ha^{-1})
'Loadel'			
March 20	113.3 ± 1.4	1681 ± 64	56.7 ± 2.0
May 18	91.9 ± 2.4	1649 ± 40	45.3 ± 1.6
'Carson'			
March 20	127.8 ± 4.7	1576 ± 74	59.4 ± 2.0
May 18	108.2 ± 2.5	1427 ± 53	46.0 ± 2.0
'Andross'			
March 21	123.6 ± 2.1	1888 ± 96	69.3 ± 2.7
May 18	115.0 ± 1.7	1766 ± 58	60.8 ± 2.7
'Ross'			
March 27	163.9 ± 7.0	1862 ± 99	80.7 ± 2.5
May 19	163.9 ± 3.2	1639 ± 69	72.2 ± 3.1

FW, fresh weight.

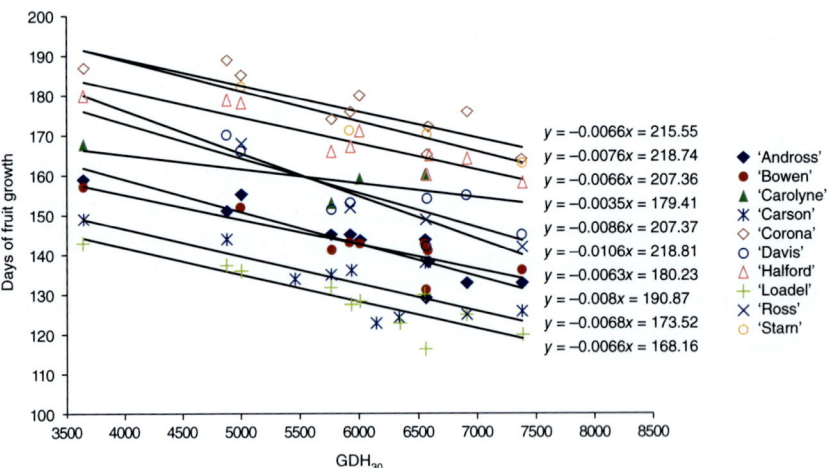

Fig. 9.15. The relationship between temperature accumulation (growing degree-hours) during the first 30 days after bloom (GDH$_{30}$) and the number of days of fruit growth for ten different clingstone peach varieties grown in California. Note that the days of total fruit growth can vary by about 25 days depending on spring temperatures and that the relationship is similar for early- and late-season varieties. (From Ben Mimoun and DeJong, 1999.)

Understanding the carbohydrate economy of fruit trees can provide answers for orchard outcomes. During warm springs the tree must provide more resources in a shorter time period to individual fruit and the tree simply can't supply resources fast enough to keep up with resource demand, compared to cooler springs. As other fruit RGR analysis has shown, carbon deficits early in the season that result in reductions in fruit growth relative to their potential cannot be made up later in the season, so final fruit sizes tend to be smaller than normal. Furthermore, it is likely that the carbon deficits that limit fruit growth early in the season may be due to transport limitations as well as assimilate availability limitations. If this is the case, then in extreme cases, even early heavy fruit thinning cannot mitigate for smaller fruit sizes that are linked to high early spring temperatures. For this reason, growers of temperate fruit in warm, subtropical regions often experience small fruit sizes that cannot be substantially improved with early fruit thinning.

In California, knowledge of the dependence of the length of the fruit development period and fruit size on early spring temperatures has been used to provide growers with a web-based decision support tool. If growers keep track of the date of full bloom for an orchard, they can go to the weather page of the UC Davis Fruit and Nut Research and Information Center website (http://fruitsandnuts.ucdavis.edu, accessed 20 August 2021) 30 days after bloom and obtain data regarding the accumulated growing degree-hours from the weather station nearest to their orchard. From this they can predict how long the fruit development period is going to be and adjust management practices accordingly. In especially warm spring years growers should try to thin their fruit crops as soon as is practically feasible to minimize fruit size problems.

The relationship between spring temperatures and fruit development/growth pertains to most temperate fruit species. In addition to peach and nectarine, data have been reported for plum, prune, almond, cherry and pear. There is also some evidence with apple, peach and almond indicating that the post-fertilization abortion of immature fruitlets (sometimes referred to as "June drop") that occurs fairly early in fruit development is triggered by a lack of assimilates. This abortion appears to occur during a critical period when carbohydrate availability mobilized from storage becomes limiting and rapidly increasing fruit growth becomes dependent on limited assimilates from new leaves. It is important to understand that this "June drop" phenomenon is not caused by the "tree dropping its fruit" but by the abortion of individual fruits that apparently aren't supplied the minimal amounts of resources necessary to continue development and growth.

If global climate change brings an increase in spring temperatures, then one of the most pronounced effects is likely to be a shortening of fruit development periods. If high spring temperatures become the norm, this could lead to a tendency toward reduced fruit size and corresponding decreases in yield of individual fresh market fruit crops. Similarly, this could lead to increased immature fruit drop ("June drop") and reduced yields due to decreased final crop loads in other species like almonds and prunes.

Nut Development and Growth

The principles governing the development and growth of fruit on nut trees are similar to fruit trees. However, since producing the seed in the fruit of nut trees is the principal goal of nut production, it is important to focus on seed development and growth in nut tree fruit. The overall fruit dimensions of almond, walnut and pistachio fruit increase rapidly after fruit set and approach a plateau after about two months. However, kernel development within the fruit goes through several stages as the fruit matures. In almonds and walnuts, the seed cavity in the fruit is initially filled with an immature kernel filled with translucent nucellar tissue and a developing seedcoat. Over time the nucellar tissue in the developing kernel is replaced by endosperm that is later absorbed by the developing cotyledons of the embryo (Figs 9.16 and 9.17). Initially the developing kernel is high in sugars but as the kernels mature, sugars decrease and are replaced by proteins and fatty acids (lipids) (Fig. 9.18). Interestingly, this process represents an increasing progression of energy investment in the kernel from rather "low-cost" nucellar tissue to "higher-cost" proteins and fatty acids.

Pistachio fruit development is quite different from other nuts. The final size of pistachio fruit is attained rapidly after fruit set but initially only consists of a mostly hollow shell covered with a hull (pericarp) and a very small embryo inside of it (Fig. 9.19). The embryo grows and fills the hollow shell later until it usually splits the shell just before fruit maturity. Similar to almonds and walnuts, the kernel is initially high in sugars but they are replaced by more energy-intensive lipids as the fruit matures (Fig. 9.20).

Alternate Bearing

Alternate bearing is the condition when there are large annual swings in crop productivity such that in extreme cases, over a two-year production period, 70 to 100% of the production may occur in one year and 0 to 30% of the production occurs in the subsequent year. Alternate bearing has traditionally often been thought to be a mechanism that is built into the behavior of trees so that trees can "recover" from the stress of overcropping in a particular year. Accordingly, it is hypothesized that tree survival is not compromised by bearing excessively large crops in two successive years. Thus, alternate bearing (Fig. 9.21) is often thought of as being controlled at the whole-tree level.

However, there is growing evidence that alternate bearing in many temperate fruit crops is mainly controlled at the shoot or spur level and alternate bearing at the tree or orchard level is mainly a consequence of synchronization of the flowering and fruit bearing of the collective population of bearing shoots or spurs in a tree and orchard. It has been clearly shown that individual spurs (in apple, pear, almond, walnut) or shoots (pistachio) rarely flower in the year after they bear fruit. The cause for the inhibition of spur or shoot flowering after

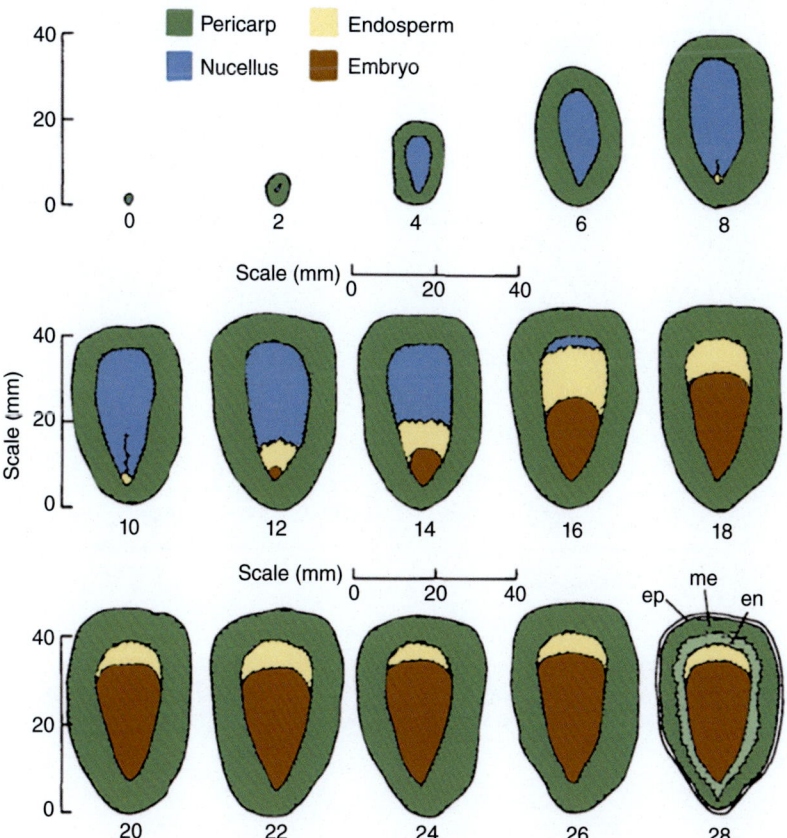

Fig. 9.16. Diagrammatic representation of almond fruit growth. Note that the fruit and kernel achieve near-maximum size by eight weeks after bloom. The kernel is first filled with "cheap" nucellar tissue and then mostly with "more expensive" endosperm tissue and finally with the embryo that has increasing concentrations of fatty acids and proteins. When mature, the epicarp (ep), mesocarp (me) and endocarp (en) are clearly distinguishable in the fruit. (Adapted from Hawker and Buttrose, 1980.)

bearing fruit has been a topic of much research. It is important to remember that because of the timing of fruit and flower development in most deciduous fruit trees, floral development for the next season's crop often overlaps with fruit development and growth in the current season.

Thus, the presence of fruit on a spur or shoot in one year can inhibit the initiation and/or development of flowers or inflorescences on the same spur or shoot. In almond, the leaf area of fruiting spurs is reduced, and reduced spur leaf area has been linked to lack of spur flower production (Fig. 9.22). Some researchers believe that this is a matter of limited resource availability to support the simultaneous development of both fruit and flowers, while others have

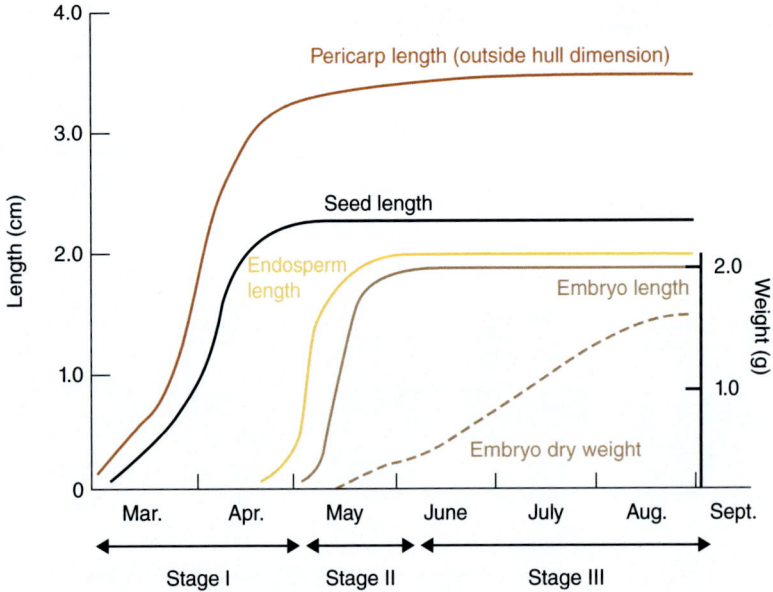

Fig. 9.17. Pattern of almond fruit and kernel growth. Length growth of the embryo (in Stage II) lags behind overall seed length growth (Stage I) and is followed by slower increases in embryo dry weight (during Stage III). (From J.M. Labavitch and V.S. Polito, UC Davis, 2014, personal communication.)

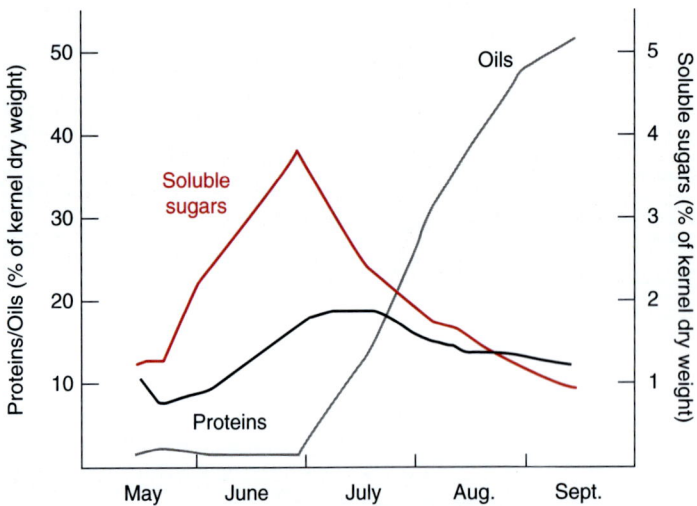

Fig. 9.18. Changes in proportions of fats (oils), proteins and soluble sugars in 'Ashley' walnut kernels during development. (From Pinney *et al.*, 1998.)

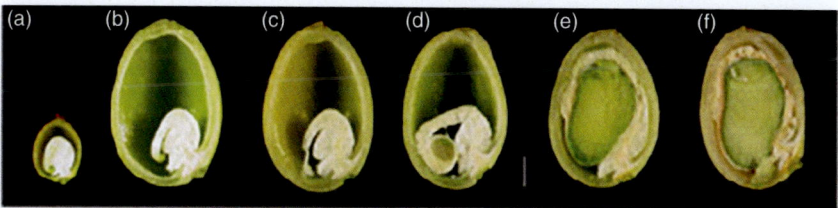

Fig. 9.19. Pistachio nut development is very different from other nut species. A hollow shell is developed first and then the embryo (kernel) grows and usually outgrows and splits the shell. (From L. Ferguson, UC Davis, 2012, personal communication.)

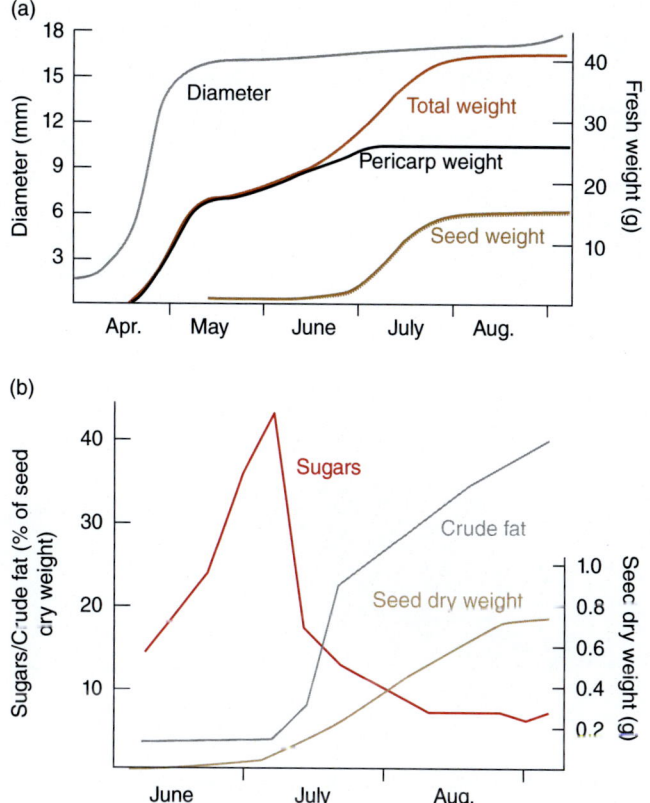

Fig. 9.20. The overall dimensions of pistachio fruit grow very rapidly and maximum seed weight is achieved at least three months later (a). Sugar concentrations peak early, followed by accumulation of lipids (fat) (b). (From J.M. Labavitch, UC Davis, 2012, personal communication.)

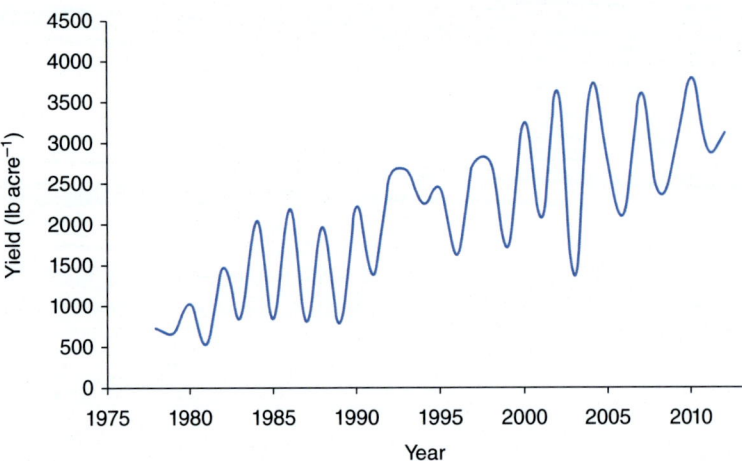

Fig. 9.21. Pattern of average California pistachio yields per acre from 1978 to 2013. Note the strong swings indicative of significant alternate bearing. (From Geisseler and Horwath, 2016.)

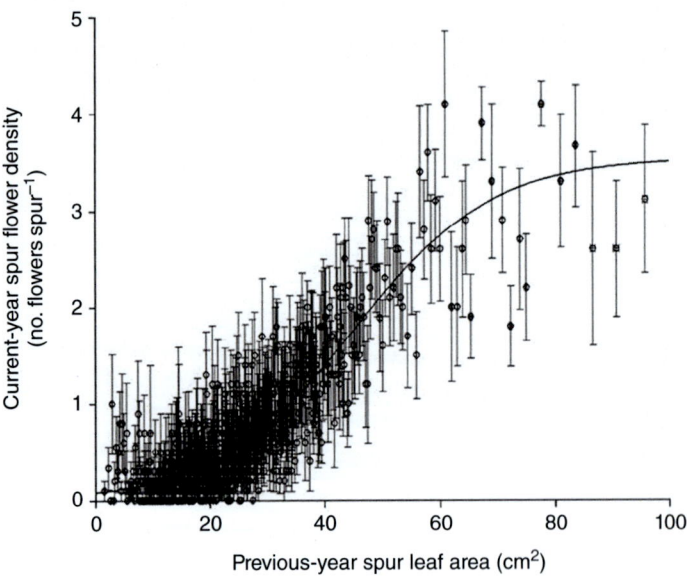

Fig. 9.22. The relationship between almond spur flowering and previous-year spur leaf area. Almond spurs rarely produce two years in a row because fruit production on a spur decreases leaf area on that spur and spurs with small amounts of leaf area in one year produce few flowers in the subsequent year. Each point is the mean of ten spurs with vertical bars indicating standard error of the means. (From Tombesi *et al.*, 2016.)

focused on hormonal signals coming from the fruit that may influence floral initiation or development. Regardless of the signaling mechanisms involved, alternate bearing at the spur or shoot level can lead to alternate bearing at the tree or orchard level when most of the bearing units on a tree become synchronized. Synchronization usually occurs because of environmental events such as frost or inclement weather near bloom time that may limit fruit set in one year and, thus, a majority of bearing units are set up to flower in the subsequent year. Alternatively, very favorable conditions that support excessive fruit set in one year can lead to decreased floral development for the next year.

The primary management tools for mitigating alternate are fruit thinning and pruning. Judicious fruit thinning can reduce or eliminate flowers on some spurs/shoots and thus floral development can occur on those spur/shoots for bearing the subsequent year. However, the fruit thinning must be done early enough so that immature fruits do not inhibit the initiation of floral development before they are thinned. The timing of effective fruit thinning can vary by cultivar within a species. Pruning can mitigate alternate bearing because trees respond to pruning by producing new shoots and, thus, regular pruning ensures a heterogeneous age structure in the population of bearing units and decreases flowering synchrony among the bearing units.

Understanding the Long-Term Storage Sink

10

Similar to short-term starch storage in the chloroplasts of the leaves that serves to buffer growth of organs from carbohydrate shortages due to diurnal patterns of photosynthesis related to daily patterns of light and darkness, trees also have long-term storage capacity to enable them to supply the minimal respiratory needs of tissues during the winter and resume growth in the spring when trees are still leafless. This long-term storage of carbohydrates and some minerals occurs primarily in the phloem and xylem tissue of the branches, trunk and roots. While active phloem tissue has higher concentrations of stored carbohydrates than xylem tissue, the mass of active xylem storage tissue is many times the mass of the active phloem tissue. Thus, xylem tissue comprises the largest storage compartment of temperate deciduous fruit trees.

The cells involved in carbohydrate storage in xylem are the xylem parenchyma cells and they actively store carbohydrates in the summer and fall months and remobilize them in the spring (Fig. 10.1). Xylem parenchyma cells can apparently maintain their active carbohydrate storage/remobilization activity for several years and all sapwood probably serves this function (Fig. 10.2). Soluble carbohydrates (mainly sucrose and sorbitol in trees in the Rose family) are transported to the xylem storage cells through the phloem and ray parenchyma cells. There, the soluble carbohydrates are converted into insoluble starch. However, during the winter trees maintain a relatively constant concentration of soluble carbohydrates in their woody tissues by continuously synthesizing and degrading starch in response to ambient day/night temperatures. This helps to maintain solute concentrations in these cells and acts as a sort of antifreeze that protects tissue integrity during cold weather. In spring the stored starch that is degraded to sugar is mobilized and "loaded" into xylem vessels adjacent to the xylem parenchyma and transported up the tree due to "root pressure" generated by the solutes in the xylem vessels or carried in the transpiration stream as the buds in the canopy begin to grow and transpire.

Although many researchers have considered long-term storage of carbohydrates to be a passive process, there is growing evidence that long-term storage is an active sink for assimilates and it is logical that the total xylem sink

© T.M. DeJong 2022. *Concepts for Understanding Fruit Trees* (T.M. DeJong)
DOI: 10.1079/9781800620865.0010

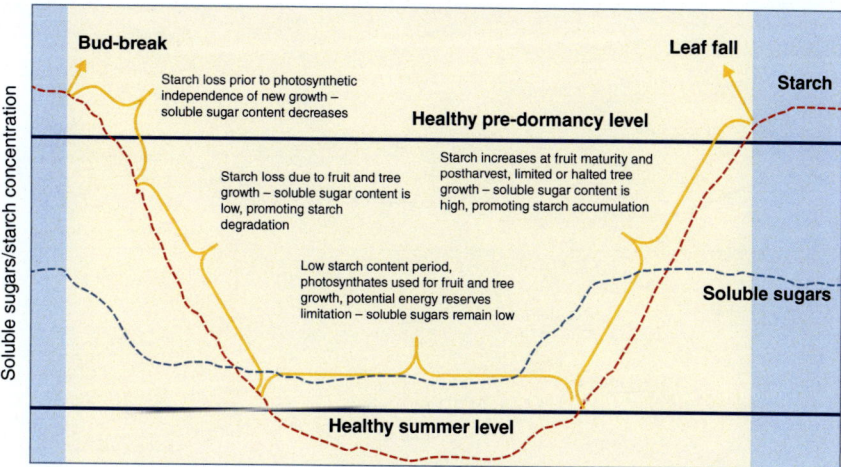

Fig. 10.1. Fruit tree non-structural carbohydrate concentrations in woody tissues are high in the late fall and winter and decrease rapidly as trees come out of dormancy in spring. Carbohydrates reserves are built up again in late summer and fall before dormancy. (From Zwieniecki and Davidson, n.d.)

storage capacity is mainly constrained by the anatomical development and growth of woody tissue. Since the number and volume of xylem parenchyma cells are determined at the time an annual ring of xylem is formed, the potential collective mass of all xylem parenchyma cells older than the current season is fixed and only the increment of xylem storage added in the current season is subject to the conditions of the current season. Thus, if radial growth of a tree is limited due to environmental stress in a given season, that can affect the incremental growth of storage capacity for that season and also limit the contribution of that year's xylem storage capacity in subsequent years.

Since the majority of long-term non-structural carbohydrate storage occurs in the xylem tissue, the tree trunk, large branches and large roots are the main storage organs of the tree. Even though roots generally account for less than one-third of the total vegetative biomass of a tree, tree roots can account for as much as one-half of the stored carbohydrates in a tree because the non-structural carbohydrate concentration in the roots can be twice that of the branches. During the dormant season a large old apple tree has been reported to contain the equivalent of 11.2 kg (~25 lb) of sugar stored in its roots and a little more than that in its trunk and branches (Table 10.1).

At the beginning of the dormant season the storage in the sapwood is at its annual peak and the majority of that is in the form or starch. As the seasons progress the trees go dormant and metabolic activity in the tissues is very low, but a minimal amount of cell activity continues, and this activity is supported by stored carbohydrates. In trees growing in northern latitudes, some of the stored starch is converted to soluble sugars and this decreases the osmotic

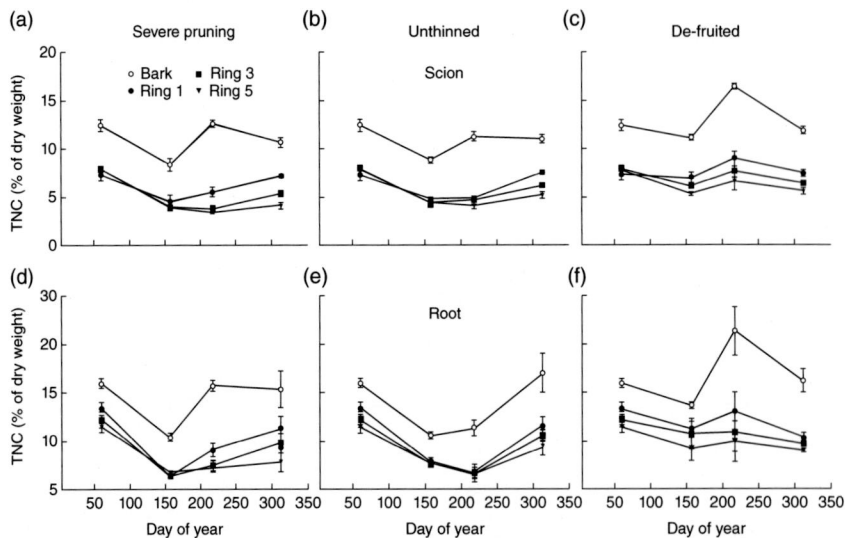

Fig. 10.2. Seasonal patterns of total non-structural carbohydrates (TNC, % of dry weight) in the active bark and xylem tissues in the one-, three- and five-year-old tree rings of the trunk (scion) (a, b and c) and root crown (d, e and f) of peach trees subjected to severe pruning that forced strong vegetative growth (a, d), very heavy crops (b, d) or no crop (c, f). Symbols indicate mean values with vertical bars indicating standard errors of the means. (From Da Silva *et al.*, 2014.)

Table 10.1. Distribution of biomass and stored sugar and starch in an 18-year-old, large apple tree growing in Missouri in mid-October. (Data adapted from Murneek, 1942.)

	Dry weight (kg)	Starch and sugar (kg)	% of oven-dry weight
Leaves	14.89	1.40	9.41
Spurs	2.87	0.31	10.90
Wood aged:			
1–3 years	9.89	1.01	10.31
4–6 years	21.26	1.12	5.36
7–10 years	56.25	3.26	5.81
11–18 years	45.90	2.70	5.88
Main stem	30.93	3.43	11.10
Aboveground	**181.99**	**13.23**	**7.29**[a]
Root stump	28.49	4.51	15.83
Roots aged:			
1–6 years	2.45	0.50	20.37
7–13 years	10.21	2.50	25.51
14–18 years	21.27	3.73	17.53
Belowground	**62.42**	**11.24**	**18.01**[a]
Total for tree	**244.41**	**24.47**	**10.03**[a]

[a]Weighted means relative to the weights of the individual plant parts.

potential of cells and functions as a type of "antifreeze" that protects against ice formation in tissues and protects cells from rupturing. During this period some trees such as sugar maple and walnuts can be tapped for producing syrup. Recent research indicates that most temperate trees have cellular mechanisms that regulate xylem soluble carbohydrate concentrations on a daily basis by up- and downregulating starch synthesis and degradation enzymes in response to diurnal swings in temperature. It has been hypothesized that the end of endodormancy in spring is actually triggered by increasing temperature conditions that limit the xylem tissue ability to regulate concentrations of soluble carbohydrates. Thus the storage sink may be instrumental in initiating seasonal growth activity.

The spring flushes of flowering, shoot and root growth require substantial amounts of carbohydrates to support growth and respiration of these organs. Thus, stored carbohydrates decline rapidly in the spring and early summer. Stored carbohydrates are usually at a minimum by early summer and their recovery is dependent on the phenology and amount of cropping. If cropping is minimal or in the case of trees with fruit that matures early in the growing season, the storage sink can begin to recover during the summer months. If crops are heavy and continue to be actively growing until late summer or fall, they can compete with the storage sinks for carbohydrates and recovery of stored carbohydrates is mainly dependent on photosynthesis that occurs in the postharvest period. Under these conditions it is particularly important to maintain trees in good health until dormancy occurs.

All xylem sapwood appears to be active in storage and mobilization of carbohydrates. Efficiency of these processes may decline with sapwood age and outer rings of xylem appear to be somewhat more active than the inner rings. Although all annual rings of sapwood appear to be active in non-structural carbohydrate storage as the tree prepares for dormancy in the fall and during mobilization to support regrowth in the spring, outer (younger) rings appear to fill earlier in the fall and release carbohydrates earlier in the spring compared to inner rings. Thus, carbohydrate storage/mobilization efficiency may decrease with ring age. Eventually, inner rings of xylem are no longer active in storage when they become fully lignified and form heartwood.

Integration of Tree Source and Sink Activities

11

Plant physiologists and agronomists have been building computer models to simulate or predict growth and productivity of various crops for more than 50 years. Some such exercises have been very successful, particularly with annual crops such as maize, cotton and tomato. Annual crop models are often initiated with a germinated seed or small plant and then simulated growth of the crop occurs based on algorithms derived from analyzing crop growth over time under specific growth conditions. As the simulated crop matures it flowers and sets fruit, and yields can be predicted fairly well based on simulated accumulated biomass that is converted into harvested product. Annual crop models have been very useful tools for identifying key factors that influence and/or limit crop growth and productivity, facilitating development of best management strategies, and predicting optimal dates of crop maturity for managing harvest logistics and postharvest processing.

Modeling of perennial tree fruit crops is much more complex and predicting yields is much more difficult because of several factors. First, tree crop growth occurs over multiple years and thus a proper model must be capable of simulating tree growth over multiple years. This is complicated by the fact that most fruit and nut trees are not allowed to grow entirely naturally but are manipulated (trained/pruned) with human intervention. Such interventions can be arbitrary, and the combination of biological variations and human interventions results in every tree in an orchard being uniquely different from all other trees in an orchard. Thus, validation of tree models is more difficult than for annual crop plants. Modeling growth over multiple years additionally requires modeling not only of canopy photosynthetic activity and distribution of photosynthates to support the growth of various organs of a tree within a growing season, but also of carbohydrate storage through the dormant season and remobilization of stored carbohydrates to support regrowth the following spring. Furthermore, because flower development in many fruit species begins in the year prior to actual anthesis of the same flowers, factors influencing flowering and fruit set occur in two growing seasons. Anthesis in most temperate deciduous tree crops also occurs in late winter or early spring, a period

© T.M. DeJong 2022. *Concepts for Understanding Fruit Trees* (T.M. DeJong)
DOI: 10.1079/9781800620865.0011

when weather is very unpredictable, whereas anthesis in many annual crops occurs in summer when temperatures tend to be relatively more predictable.

However, probably the most important factor inhibiting the development of perennial tree fruit crop models was the fact that traditional annual crop models used deterministic, empirically derived partitioning coefficients to distribute dry matter derived from estimates of canopy photosynthesis over time. Partitioning coefficients were derived by growing crops under various environmental conditions and then harvesting entire plants and analyzing the mass of each part of the plant at time intervals as the crop grew. This approach to modeling photosynthate partitioning over the life of perennial crops was not feasible for trees. For these reasons, crop modeling and integrated understanding of crop physiology of tree fruit crops have lagged behind annual crops. Nevertheless, it was clear that tree crop modeling could be instrumental in facilitating integration of numerous aspects of the development, growth and physiology of fruit tree crops and provide a valuable tool for testing concepts for understanding how fruit trees work, if it could be achieved.

Over the past 30 years I have been involved in the process of trying to create such a model. First, we developed an integrated model of the development, growth and physiology of peach trees and then the model was extended to almond trees to demonstrate that our modeling approach could be applied to multiple tree crops. This modeling exercise has led to the discovery that, while the factors that are involved in determining the crop growth and yield of temperate deciduous fruit trees are exceedingly complex and relatively unpredictable, the actual processes involved are elegantly straightforward. The following is a synopsis of how modeling of fruit trees was approached.

As laid out in this book, a tree can be viewed economically as having a supply side and a demand side. On the supply side, the currency for life (carbohydrates) is supplied by canopy photosynthesis, which is a direct function of canopy light interception. Water and mineral nutrients, which are also needed to build, maintain and provide a medium for carrying out photosynthesis and other cellular functions throughout the tree, are supplied by uptake from the soil by the roots. On the demand side, carbohydrates and nutrients get distributed to sinks (semi-autonomous organs) according to their needs to fulfill their growth potentials. Maintenance requirements are dictated by organ biological activity, which is governed by phenological patterns of development and prevailing environmental conditions.

Since fruits have long been considered a dominant sink for carbohydrates in mature bearing fruit trees like peach and apple, the initial modeling work for defining the demand side of the carbon economy of peach trees was focused on describing the carbohydrate requirements of peach fruit over time. This work involved the measurement of fruit respiration and dry matter accumulation of peach fruit on modestly cropped trees in the field over the growing season. Since respiration rates were calculated per unit of fruit dry mass, growth was also expressed as a function of fruit dry mass by calculating fruit RGRs. This exercise resulted in two important discoveries. The classic double-sigmoid curve of peach fruit growth, when expressed as dry mass accumulation, is simply a mathematical result of a logarithmically descending RGR in the early phases of fruit growth, followed by a relatively stable phase of RGRs (Fig. 11.1). And fruit respiration rates are linearly related to fruit RGRs (Fig. 11.2).

After analyzing fruit growth in different ways, it became apparent that relative fruit growth rate curves, derived by measuring fruit growth on lightly cropped, non-stressed trees, could be used to estimate the potential

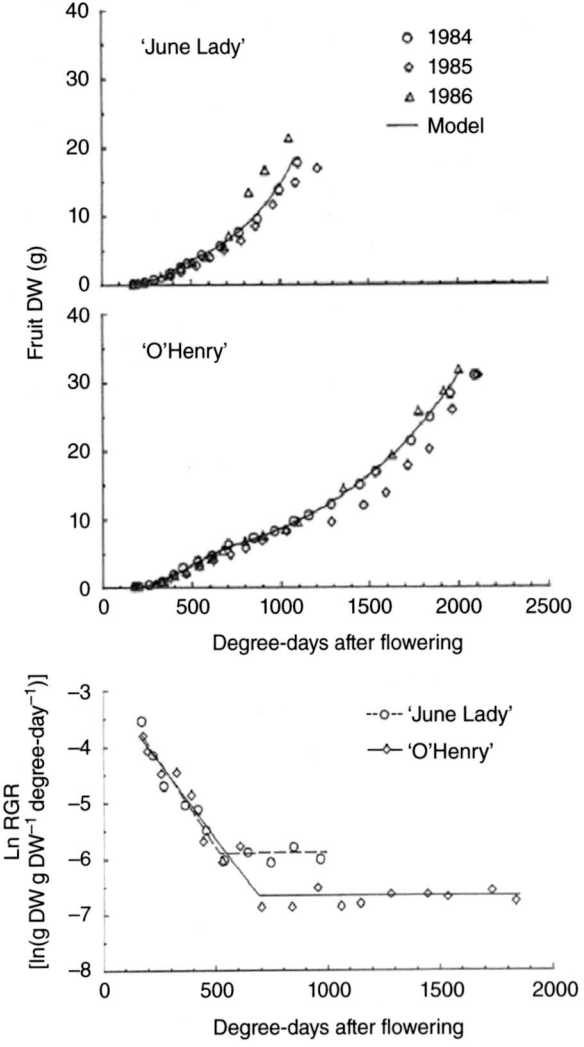

Fig. 11.1. The cumulative dry weight (DW) curve for peach fruit growth can be efficiently modeled as a rapidly descending relative growth rate (RGR) that tends to level off toward the end of fruit maturity when fruits grow exponentially. (From DeJong and Goudriaan, 1989.)

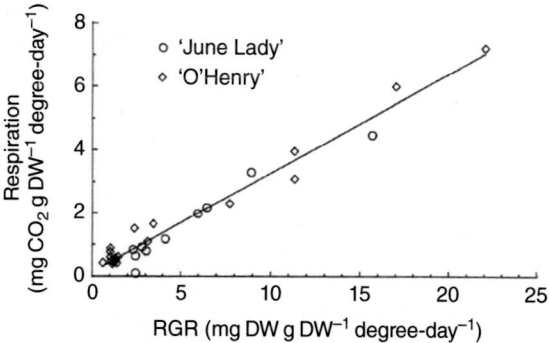

Fig. 11.2. Peach fruit respiration rates are linearly related to fruit relative growth rates (RGRs) indicating the close relationship between growth and respiration (DW, dry weight). (From DeJong and Goudriaan, 1989.)

sink capacity of fruit on a cropping peach tree. Thus a demand function for the fruiting component of a mature peach tree could be defined at specified crop loads. Demand functions for the aboveground vegetative parts of mature bearing peach trees were subsequently calculated by doing destructive harvests and measuring respiration rates of specific organs (shoots, leaves, branches and trunks). This, coupled with a classical canopy photosynthesis model based on canopy light interception, resulted in the development of a mechanistic, compartmental model of mature peach tree carbon partitioning over a growing season. The model was termed a compartmental model because carbohydrates were only distributed to the collective compartments of fruits, leaves, stems and large branches, and the trunk according to their relative demand functions as the season progressed. Roots were only given carbohydrates when the demands of all of the other organs were fulfilled (Fig. 11.3). This model demonstrated that carbohydrate partitioning in trees could be modeled without deterministic, empirically derived, partitioning coefficients (Fig. 11.4) and was useful for indicating periods of the growing season when calculated photosynthetic assimilation was not adequate to supply calculated carbohydrate demands of growing organs. It also was useful in demonstrating effects of crop load on tree growth (Fig. 11.5) and overall estimated seasonal patterns of carbohydrate assimilation and distribution to growth and respiration (Fig. 11.6).

However, the initial PEACH model was limited to simulating one growing season of a generic peach tree and fruit size versus crop load relationships were limited to estimating "average fruit size" for a specific crop load. This was not satisfactory because a major issue facing growers is that, irrespective of crop load, fruit size varies among fruits on a given tree and predicting ranges of fruit sizes at specific crop loads is critical for optimal management of crop loads (fruit thinning). Subsequent modeling efforts attempted to address this issue.

In the 1990s there were major advancements in the field of virtual plant modeling and it became possible to realistically simulate the structural development and growth of plants on computers. This made it possible to develop a

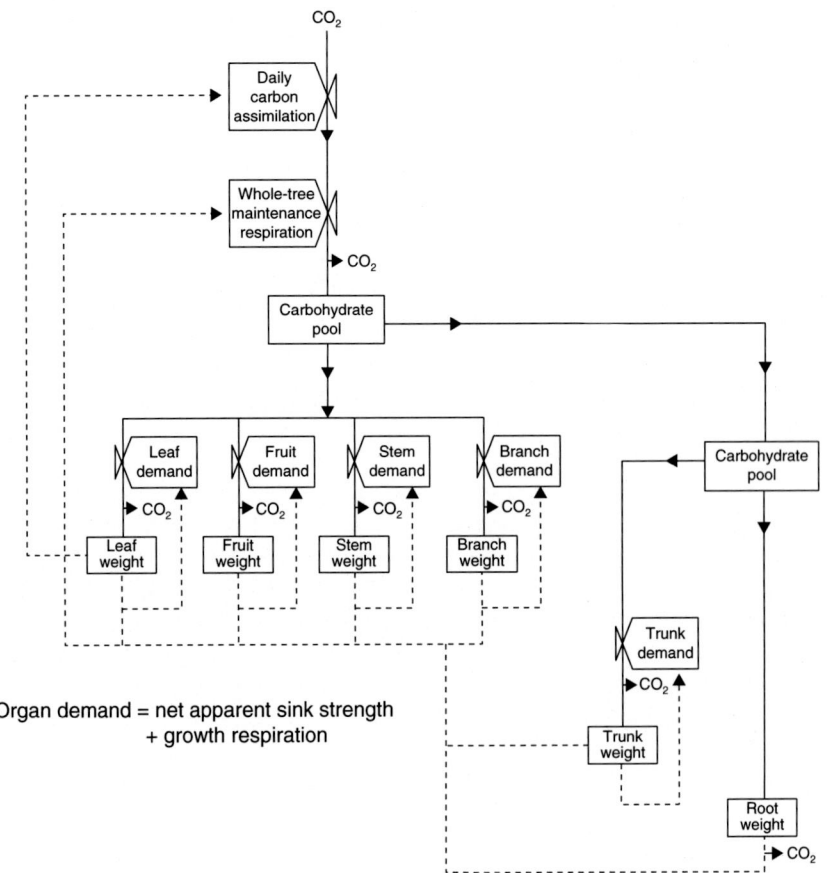

Fig. 11.3. Diagram illustrating carbohydrate distribution in the PEACH model. Boxes are state variables and valves are rate variables. Solid lines represent carbon flows and dashed lines represent information flows. In this model, daily assimilated carbohydrates are treated as a common pool that is distributed at the end of each day. (From Grossman and DeJong, 1994.)

virtual model of the aboveground structure of peach trees. The development of an L-Peach model eventually simulated the architectural development and growth of peach trees over multiple years, from a single node and leaf to a mature bearing tree (Fig. 11.7). This model not only simulated architectural tree development and growth but also incorporated calculations of physiological activity of the tree at the level of individual organs. This model is very complex and contains several components, and each component has multiple sub-models (Fig. 11.8). An architectural model was constructed using specialized L-systems software that is specifically designed to simulate the architecture of plants. Carbohydrate and water transport and source–sink dynamics were simulated using programming analogous to modeling electrical circuits.

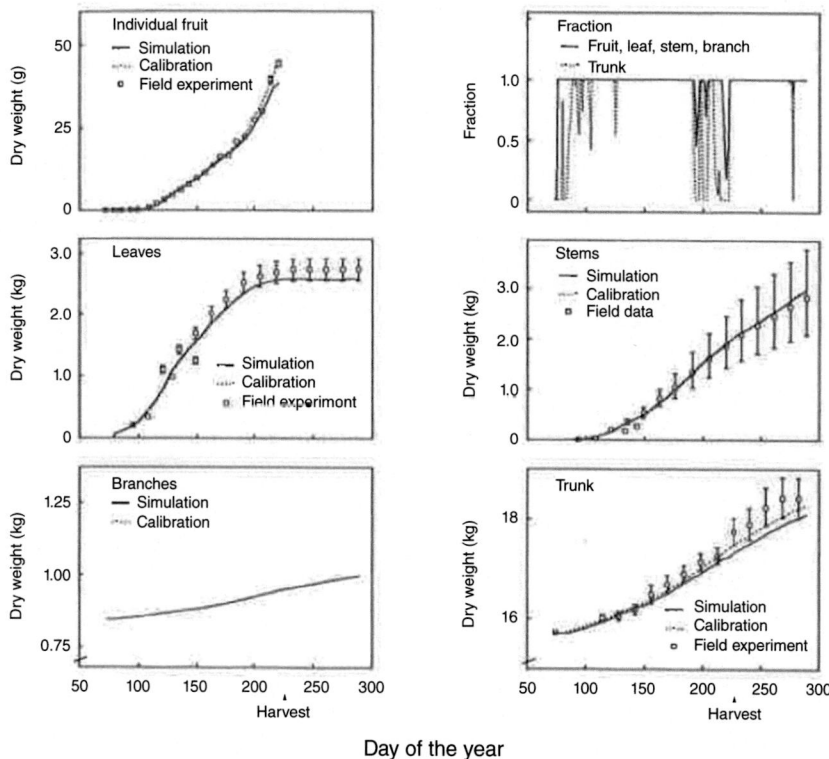

Fig. 11.4. Simulated versus actual seasonal patterns of dry weight distributions per tree under the calibration conditions for the PEACH model. Individual fruit dry weight increase and fraction of potential growth allowed by the model for trees that were heavily thinned at bloom, along with leaf, stem, branch and trunk dry weight increases for de-fruited trees. Data points indicate mean values of field data with vertical bars indicating standard errors. (From Grossman and DeJong, 1994.)

Parameters for the physiological functionality of tree organs were incorporated within sub-models for each organ type and there was a special section developed to permit simulations to be interrupted and for the virtual trees to be manually manipulated (pruned).

Photosynthetic and respiration responses of every leaf in response to leaf light exposure and temperature were calculated at hourly intervals throughout multiple growing seasons. Growth and respiration of each node of shoots were also simulated along with production of flowers and fruit as they developed over the seasons. Carbohydrate distribution throughout the virtual trees was conceptually managed in a manner similar to the original PEACH model, except that each leaf was treated as a source of photosynthates and each individual stem node and associated organs (leaves, flowers or fruit) were treated as sinks.

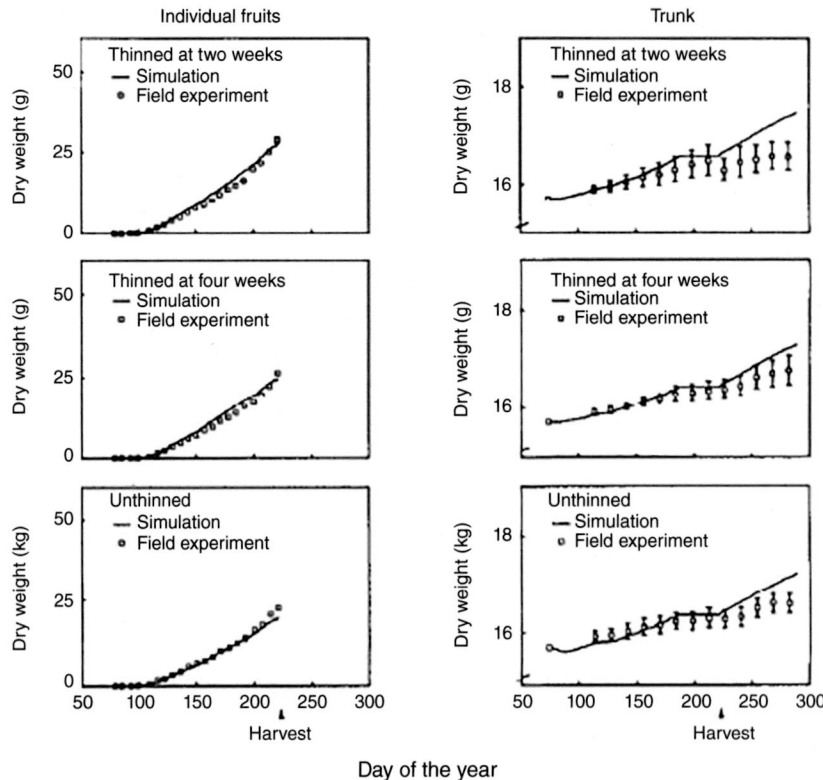

Fig. 11.5. PEACH model simulations and field data of seasonal patterns of mean peach fruit and trunk dry weight increases for trees thinned two and four weeks after bloom or left unthinned. Later thinning and non-thinning reduced achieved mean fruit size and trunk growth. (From Grossman and DeJong, 1994.)

Carbohydrate distribution within the tree was accomplished by treating the tree in a manner analogous to an electrical circuit. Photosynthesis of individual leaves was modeled as being analogous to generating an electrical charge and the flow of photosynthate was analogous to an electrical current. Thus, a circuit could be simulated with electrical potential (voltage) differences between different parts of the circuit with resistances/conductance built in at appropriate places in the circuit (Fig. 11.9 and Table 11.1). The circuitry for each organ included parallel sinks for primary growth, secondary growth, storage and photosynthesis or respiration, as appropriate. Thus, carbohydrates flowed around the circuit in a manner analogous to current in an electrical circuit. Using the electrical circuit analogy enabled the simulation of carbohydrate distribution within the tree to be modeled in the manner according to the concepts outlined in Chapter 5 of this book.

In this system transport resistances between individual sources and sinks were estimated so simulated individual fruit growth and final size varied

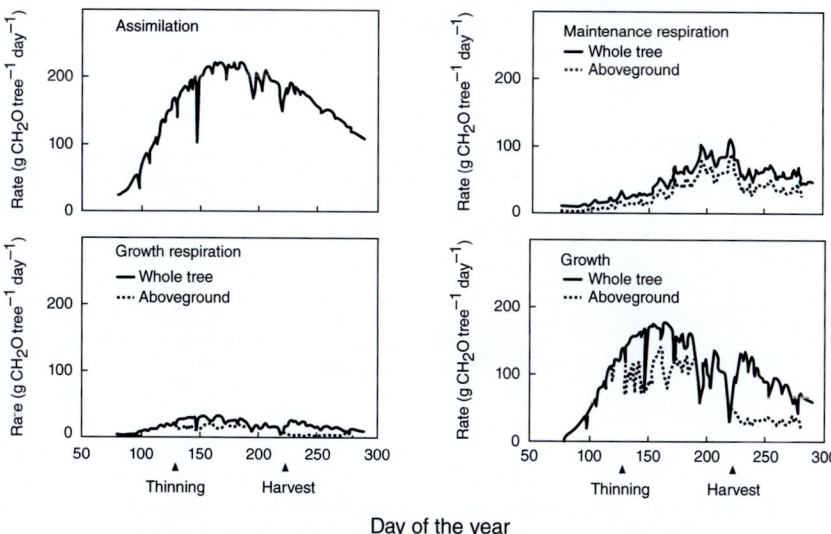

Fig. 11.6. PEACH model simulations of seasonal patterns of daily carbon assimilation, maintenance and growth respiration and growth of peach trees that were thinned eight weeks after bloom (CH_2O, carbohydrate). (From Grossman and DeJong, 1994.)

Fig. 11.7. Picture of a simulated, virtual, two-scaffold peach tree in its seventh year after planting. Virtual peach tree growth was simulated using the L-Peach model. The tree was manually pruned after each simulated year of growth in virtual space. (From simulations of the L-Peach model run by D. DaSilva.)

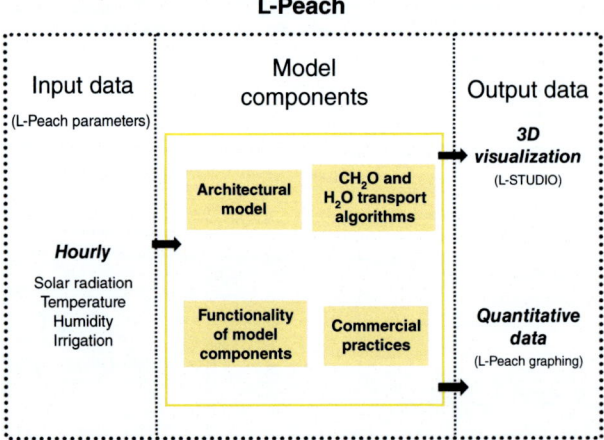

Fig. 11.8. The L-Peach model has several major components. A system for inputting actual hourly values of environmental data and adjusting model parameters. A system for displaying the visual growth of a tree as well as quantitative data for every part of the tree in hourly time steps. Internally there is an architectural component for simulating the structure and growth of each organ of the tree, carbohydrate (CH_2O) and water transport algorithms, sub-models for calculating the functioning of each organ of the tree, and a system to start and stop simulations to simulate pruning and fruit thinning operations. (Schematic by D. Da Silva.)

depending on their location relative to specific sources (leaves) and other sinks (especially fruit). Thus, this model was capable of simulating distributions of fruit sizes for specific crop loads (Fig. 11.10) as well as the effect of timing of crop load management (fruit thinning) on final fruit sizes.

The capacity to model the architectural structure of trees stimulated increased interest in the actual structure of trees. Development of algorithms for pruning virtual trees and simulating tree growth responses to pruning in turn stimulated systematic analyses and modeling of shoot growth. This led to detailed studies of bud fates along peach shoots. Later versions of the model incorporated statistical sub-models for predicting where flowers/fruit, as well as lateral branching, occur on shoots of different lengths. Such studies led to the discovery that specific shoot types (proleptic and sylleptic) of peach trees have very similar, distinct structures and are limited to specific numbers of nodes (Fig. 11.11). The capacity to incorporate the structures of different types of shoots and model how heavy pruning can elicit the growth of these shoot types created the possibility to simulate differential responses to various types of pruning (Fig. 11.12) and illustrate why pruning alone does not efficiently control tree size.

Later versions of the model also incorporated simulation of water uptake by the roots, water transport through the tree and water loss via transpiration by the leaves. Water transport within trees was simulated by using an electrical circuit analogy similar to what was done with carbohydrate transport.

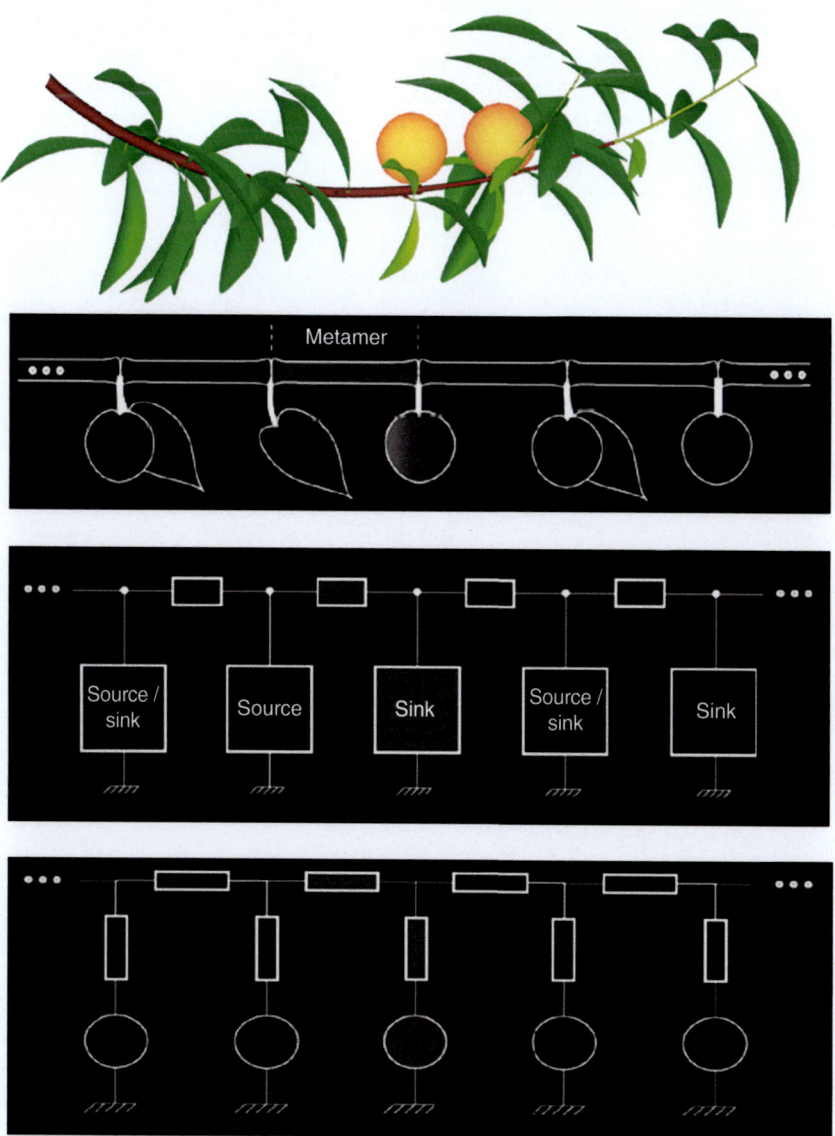

Fig. 11.9. Modeling involves simplifying things so that they can be dealt with abstractly. In the L-Peach model, source–sink relationships within shoots and the entire tree are simplified by treating each organ as a component in an electric circuit. This permitted applying concepts used for understanding electrical circuits to be used in modeling carbohydrate and water flow in trees. (Schematic by D. Da Silva.)

By estimating resistances to water movement at various key points in the soil–plant–atmosphere water pathway, as well as rates of canopy transpiration, it was possible to estimate xylem water potentials in various

Table 11.1. Each electric entity in the electrical circuit diagram depicted in Fig. 11.9 has a corresponding physiological/hydraulic meaning/entity related to the movement of carbohydrates and/or water in a simulated tree. (Table developed by D. Da Silva and T.M DeJong.)

Physiological/hydraulic entity	Electric entity	Symbol
Mass or volume	Charge	q
Mass or volume flow rate	Current	I
Hydrostatic potential, pressure	Electric potential	V
Pressure difference	Potential difference, voltage	v, e
Hydraulic resistance	Resistance	R
Hydraulic conductance	Conductance	G

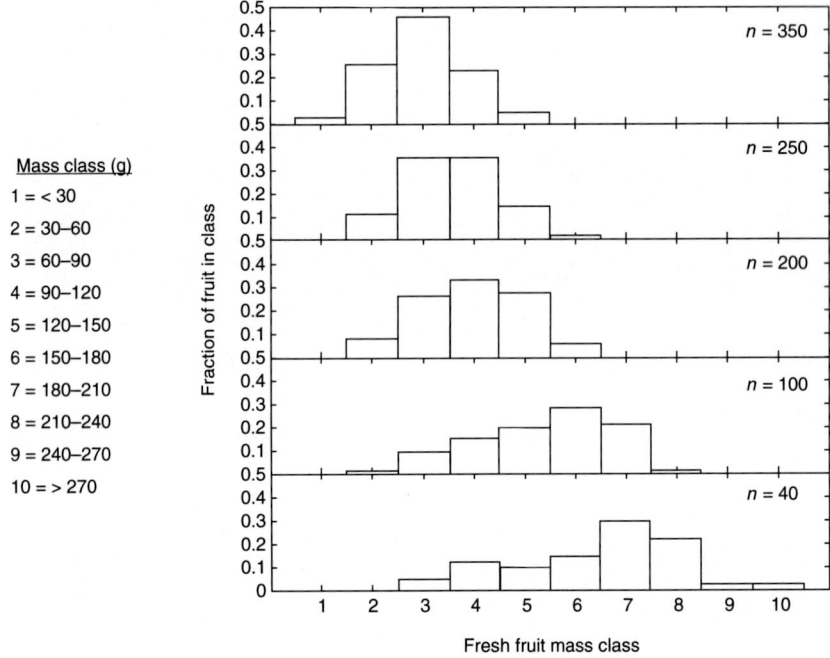

Mass class (g)

1 = < 30

2 = 30–60

3 = 60–90

4 = 90–120

5 = 120–150

6 = 150–180

7 = 180–210

8 = 210–240

9 = 240–270

10 = > 270

Fig. 11.10. L-Peach simulated effects of different crop loads (40–350 fruits tree^{-1}) on distribution of fruit size classes on peach trees with two scaffolds. Note that all trees produce fruit with a range of size classes but the number of fruits in smaller size classes increases as crop loads increase. (Simulation runs by G. Lopez and T.M. DeJong.)

parts of the tree (Fig. 11.13). This enabled simulations of the effects of irrigation intervals on tree water potential and canopy photosynthesis (not shown) that resulted in reductions in leaf, stem, root and fruit dry weight gain over a growing season (Fig. 11.14). Because the size-controlling effects of peach rootstocks are mediated by hydraulic conductance characteristics of the

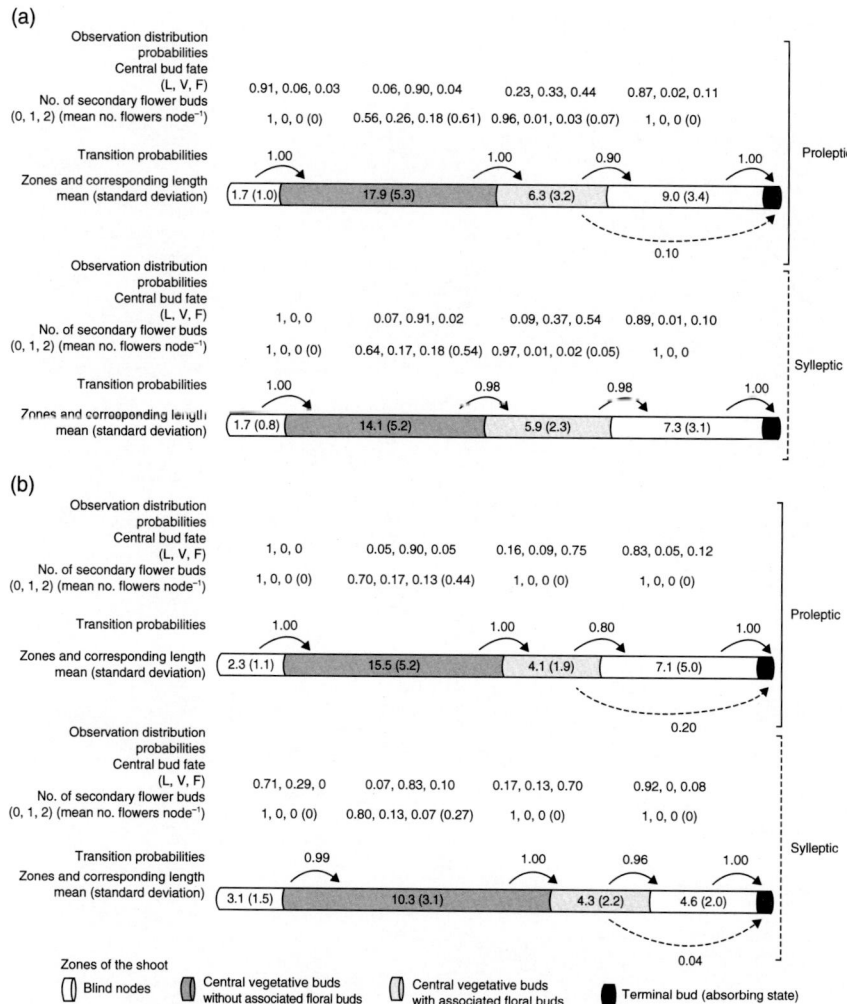

Fig. 11.11. The bud fate distribution patterns on proleptic and sylleptic shoots of 'Elegant Lady' (a) and 'O'Henry' (b) peach cultivars are very similar to one another, indicating that shoot structures on peach trees are strongly controlled by genetics. Rows of numbers indicate probabilities of corresponding central bud fates for being leaf (L), vegetative shoot (V) or floral (F) buds with secondary floral buds and the mean number of floral buds per node in each shoot zone. (From Prats-Llinàs *et al.*, 2019.)

rootstocks and their effects on shoot water potentials, modeling water transport in trees also enabled simulation of the effect of rootstocks with low hydraulic conductance on the growth of the scion of peach trees (Fig. 11.15).

It is widely recognized that long-term carbohydrate storage (mainly in the form of starch) is critical for tree survival in winter months when trees are dormant, as well as for supporting flowering and initiating vegetative growth in springtime. However, simulation of a storage sink as a separate organ in

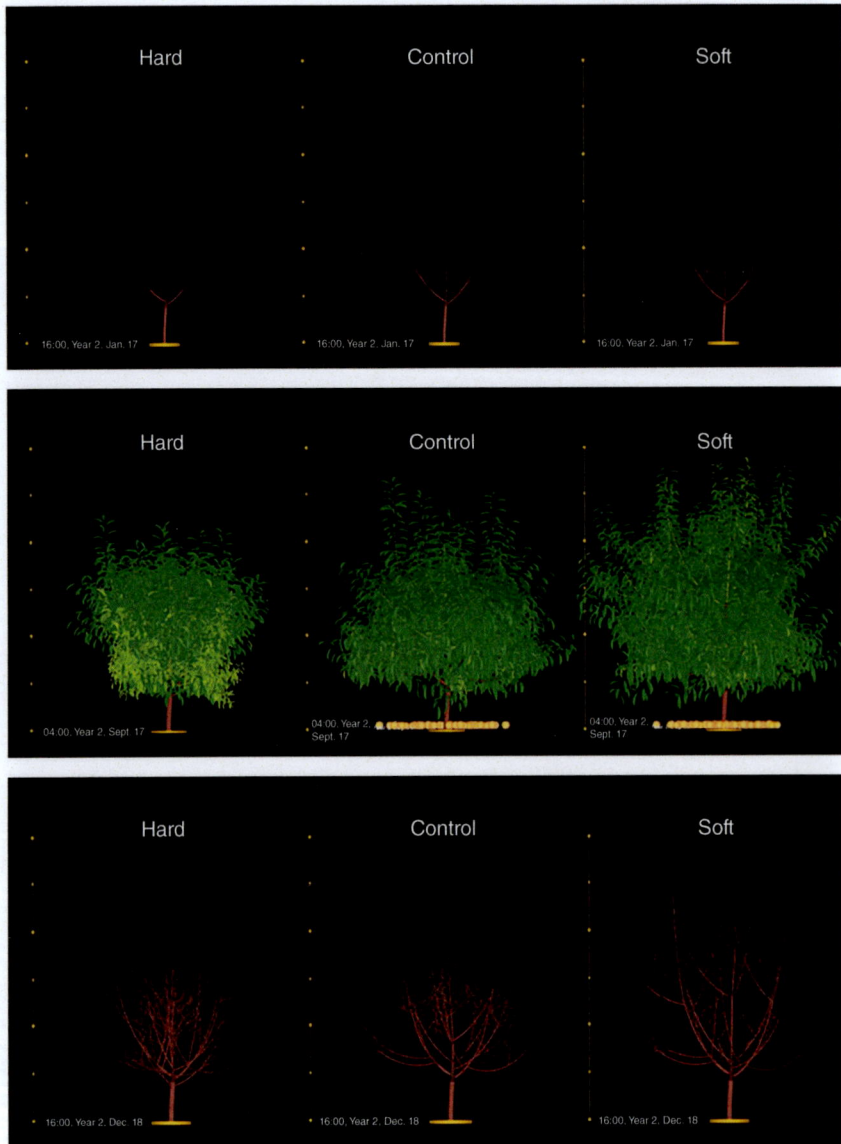

Fig. 11.12. An L-Peach set of simulations demonstrating the effects of different pruning strategies on trees after the first year after planting in an orchard. The tree on the left was short-pruned to three scaffold branches that were headed at ~10 inches above the trunk ("Hard"). In the center tree, the same scaffolds were headed at ~30 inches above the trunk and side shoots were left intact ("Control"). The tree on the right was long-pruned by tipping the three scaffold branches and leaving most side shoots ("Soft"). Notice that the left and center trees are almost the same size and the branches are more dense at the end of the year because the hard heading cuts stimulated many water sprouts. The tree on the right is larger and more open than the ones to its left because the light pruning stimulated less new growth. (Simulation runs by D. Da Silva and T.M. DeJong.)

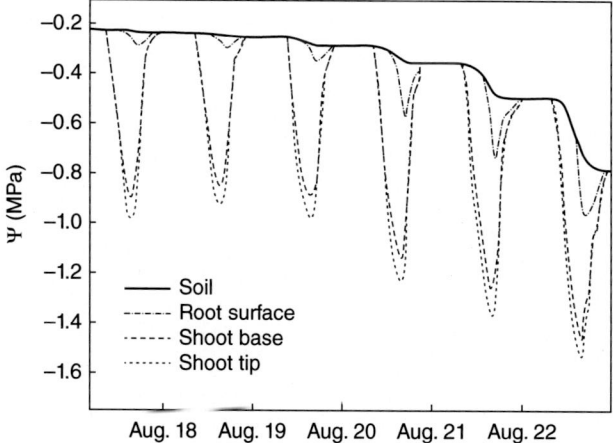

Fig. 11.13. L-Peach simulated soil, root and shoot water potentials over a six-day period as the soil begins to dry out. The difference between the curves at any specific time indicates the gradient of water potential (Ψ) that drives water movement from the soil to the shoots. (From: Da Silva *et al.*, 2011.)

the electrical circuit analogy was not possible since the majority of long-term carbohydrate storage in temperate deciduous trees occurs in the xylem parenchyma cells of sapwood. The model was capable of simulating seasonal patterns of non-structural carbohydrate concentrations in the above- and belowground woody parts of a tree (Fig. 11.16). Because the model calculated the biomass of the entire woody structure of trees, it was possible to estimate carbohydrate storage capacity and simulate the seasonal dynamics of carbohydrate storage. This enabled continuous simulation of peach tree growth over multiple years.

The development of the described model is so complex that the modeling work will never be fully completed. However, to demonstrate the utility of this modeling approach, we decided to develop an L-Almond model using the same approach. This primarily involved adjusting parameters for all of the different organs of an almond tree that were known to be substantially different than a peach tree. The architectural structure of almond trees varies much more among different almond cultivars than it does with peach trees, so virtual tree modeling of different almond cultivars involved more detailed analyses of cultivar-specific shoot structures. This research demonstrated the utility of our modeling approach to simulating the structural characteristics of almond cultivars with distinctly different tree structures (Fig. 11.17).

However, because field-grown almond trees are normally very lightly pruned compared to peach trees, simulations of these trees became very computationally heavy (slow) due to the massive numbers of elements in the circuits used to distribute carbohydrates in the simulated trees. Nevertheless, the exercise of developing the L-Almond model did demonstrate the utility of using this modeling approach for simulating the growth and physiology of another species.

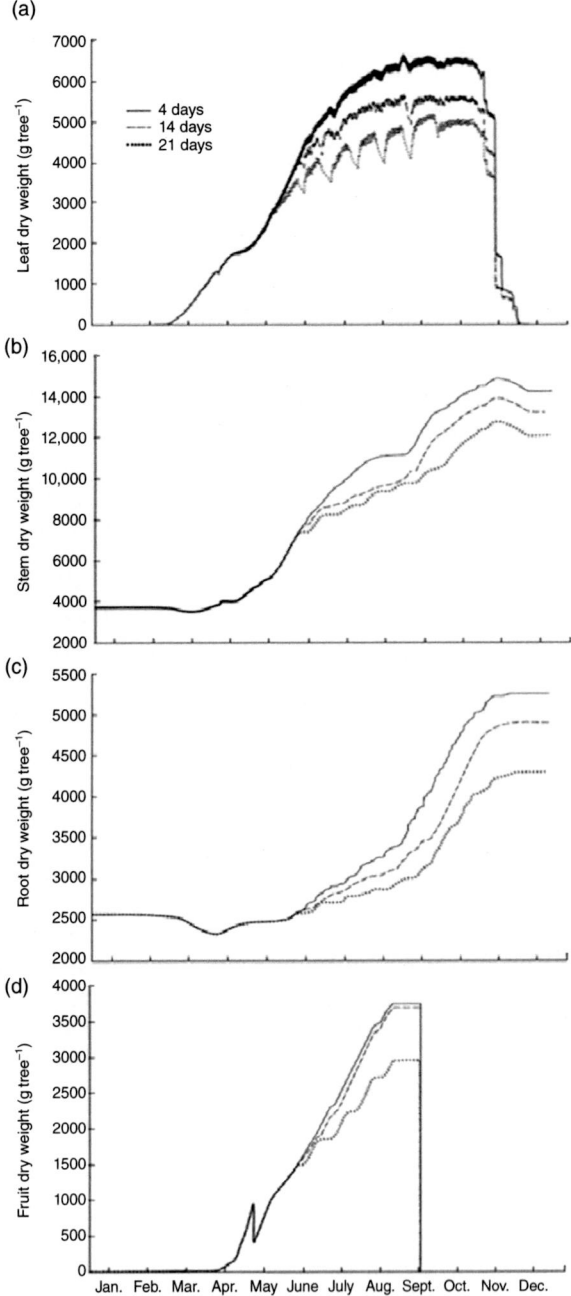

Fig. 11.14. Simulated irrigation interval (4, 14 and 21 days) effects on peach tree leaf (a), stem (b), root (c) and fruit (d) dry weight gain over a growing season. All simulated trees were confined to the same set soil volume. The early sharp drop in fruit weight indicates when the fruit were thinned. (From Da Silva *et al.*, 2011.)

Fig. 11.15. L-Peach simulated peach tree growth after four years on normal and size-controlling rootstocks. The only difference in the model parameters was a 50% decrease in hydraulic conductance of the rootstock of the tree on the right relative to the tree on the left. (From Da Silva *et al.*, 2015.)

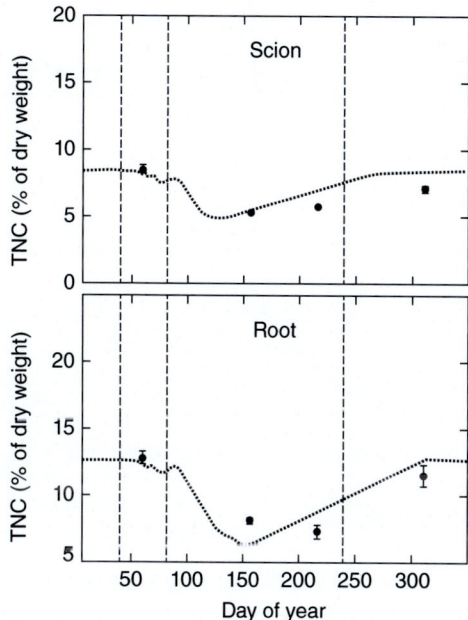

Fig. 11.16. Seasonal patterns of measured (dots) and simulated (dotted lines) mean total non-structural carbohydrate (TNC) mass fractions (% of dry weight) in the scion and root of heavily cropped peach trees. Vertical dashed lines indicate, from left to right, dates of bud-break, beginning of fruit growth and fruit harvest. Dots and associated bars indicate means and standard errors. (From Da Silva *et al.*, 2014.)

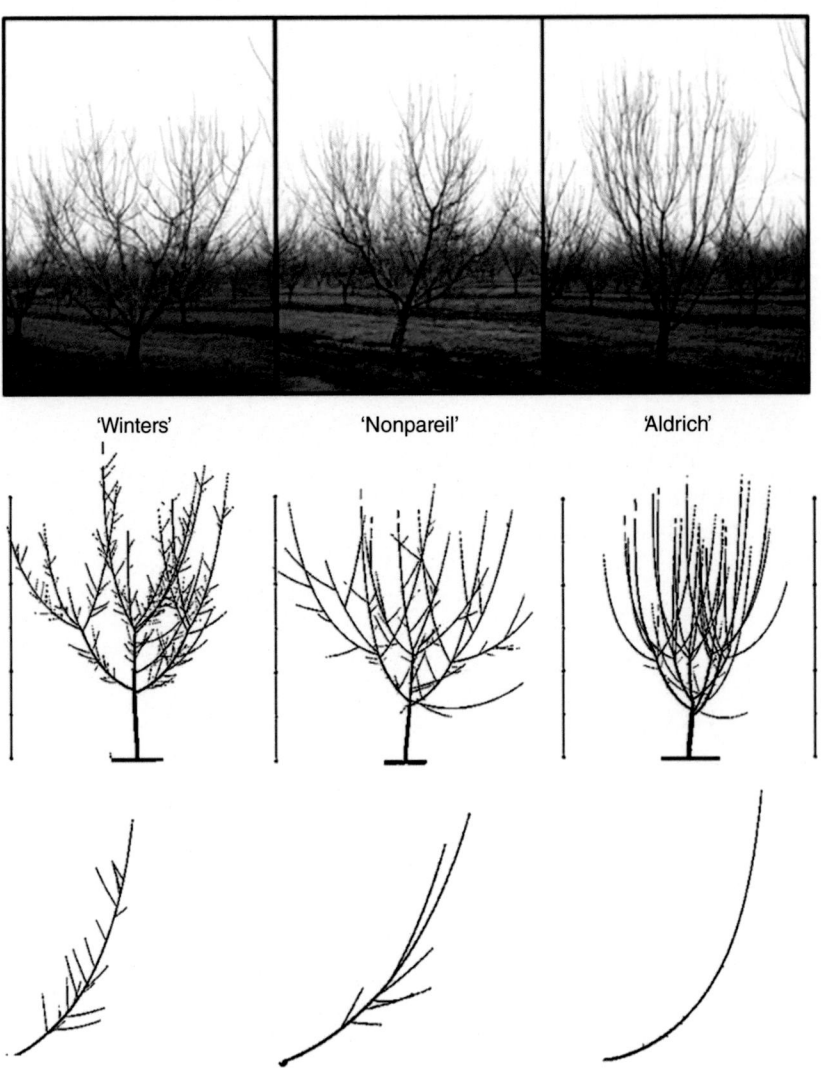

'Winters'   'Nonpareil'   'Aldrich'

Fig. 11.17. L-Almond simulations of the tree structures of three different almond cultivars resulting from the characteristic branching habits of the cultivars, compared to pictures of actual trees in the field. (From Lopez *et al.*, 2018.)

While it is not important to know the details of these modeling projects, they are very important because they demonstrate that it is possible to build an architecturally explicit, virtual, computer simulation model of tree growth and productivity based on the concepts outlined in Chapter 5 of this book. Calculating plant canopy light interception and photosynthesis, that is, the supply side of the carbon economy of trees, while being complex, has been achieved for several decades. The L-Peach and L-Almond models demonstrate that estimating the demand side of the equation is primarily a matter of developing sub-models for each of the major sinks (shoots, scaffold branches, roots, flowers and fruits, and storage sink tissues) and how their timing and activities are influenced by temperature, light and availability of nutrients and water, as well as competition from other sinks. While difficult and time-consuming, developing data to estimate parameters for all of these processes is tractable and doing the work can provide valuable insights for understanding tree growth and productivity.

The good news is that, because of the semi-autonomous nature of organs, experiments can be conducted at the organ level and data on all processes in all organs do not need to be developed simultaneously. Rudimentary models can be initially created to develop a functional modeling framework and then more detailed models can be developed with time as additional data become available. This type of modeling can be iterative, and the goal should not be to develop a final, complete model but to recognize that, because of the complexity of the system being modeled, any model will always be a work in progress and can always be improved.

Nevertheless, because such whole-plant models integrate multiple aspects of tree physiology and growth, the process of modeling can provide new insights for understanding how trees function. This book is largely an outcome of a career in fruit tree modeling.

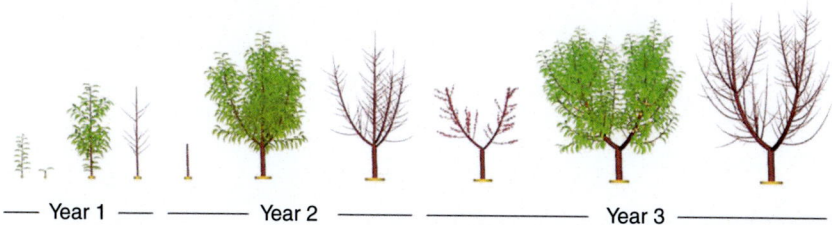

—— Year 1 —— ———— Year 2 ———— ———————— Year 3 ————————

(For access to selected dynamic simulations of the L-Peach model, see https://zenodo.org/record/47228#.YHn_uudlC70 (accessed 24 August 2021). For details of how dynamic, three-dimensional simulation modeling of peach and almond tree physiology, growth and productivity has been approached, see the various scientific publications referenced in the notes for this chapter.)

Chapter Notes and Further Reading

Chapter 1: Introduction

There are numerous books on trees and the general topic of pomology. Much of the writing regarding the structural development and growth of trees in this book was influenced by:

Wilson, B.F. (1970) *The Growing Tree*. University of Massachusetts Press, Amherst, Massachusetts.

Additional textbooks on pomology and tree fruit production that were used as general references were:

Jackson, D. (1986) *Temperate and Subtropical Fruit Production*. Butterworths of New Zealand (Ltd), Wellington.
Ryugo, K. (1988) *Fruit Culture: Its Science and Art*. Wiley, New York.
Sansavini, S., Costa, G., Gucci, R., Inglese, P., Ramina, A., *et al.* (eds) (2019) *Principles of Modern Fruit Science*. International Society for Horticultural Science, Leuven, Belgium.
Westwood, M.N. (1978) *Temperate Zone Pomology*. W.H. Freeman and Company, San Francisco, California.

In addition, ideas on the physiology and structure of trees in general were influenced by:

Cannell, M.G.R. and Jackson, J.E. (1985) *Trees as Crop Plants*. Institute of Terrestrial Ecology, UK.
Kramer, P.J. and Kozlowski, T.T. (1979) *Physiology of Woody Plants*. Academic Press, New York.
Zimmerman, M.H. and Brown, C.L. (1971) *Trees: Structure and Function*. Springer-Verlag, New York.

© T.M. DeJong 2022. *Concepts for Understanding Fruit Trees* (T.M. DeJong)
DOI: 10.1079/9781800620865.refs

Chapter 2: Energy Capture and Carbon Assimilation

Most of the concepts related to photosynthesis, respiration and plant water relations presented in this chapter are commonly described in modern plant physiology textbooks such as:

Taiz, L., Zaiger, E., Moller, I.M. and Murphy, A. (2015) *Plant Physiology and Development*. Sinauer Associates, Inc., Sunderland, Connecticut.

The approach I have taken to this topic in this book has been largely influenced by the following studies:

DeJong, T.M. and Doyle, J.F. (1985) Seasonal relationships between leaf nitrogen content (photosynthetic capacity) and leaf canopy light exposure in peach (*Prunus persica*). *Plant, Cell & Environment* 8, 701–706.
Rosati, A. and DeJong, T.M. (2003) Estimating photosynthetic radiation use efficiency using incident light and photosynthesis of individual leaves. *Annals of Botany* 91, 869–877.
Rosati, A., Esparza, G., DeJong, T.M. and Pearcy, R.W. (1999) Influence of canopy light environment and nitrogen availability on leaf photosynthetic characteristics and photosynthetic nitrogen-use efficiency of field-grown nectarine trees. *Tree Physiology* 19, 173–180.
Rosati. A., DeJong, T.M. and Esparza, G. (2002) Physiological basis for light use efficiency models. *Acta Horticulturae* 584, 89–94.

Additional references to specific studies and crops were from:

Amthor, J.S. (1989) *Respiration and Crop Productivity*. Springer-Verlag, New York.
DeJong, T.M. (1983) CO_2 assimilation characteristics of five *Prunus* tree fruit species. *Journal of the American Society for Horticultural Science* 108(2), 303–307.
DeJong, T.M. (1986a) Effects of reproductive and vegetative sink activity on leaf conductance and water potential in *Prunus persica* L. Batsch. *Scientia Horticulturae* 29, 131–137.
DeJong, T.M. (1986b) Fruit effects on photosynthesis in *Prunus persica*. *Physiologia Plantarum* 66, 149–153.
DeJong, T.M. and Moing, A. (2008) Carbon assimilation, partitioning and budget modelling. In: Layne, D.R. and Bassi, D. (eds) *The Peach: Botany, Production and Uses*. CAB International, Wallingford, UK, pp. 224–263.
DeJong, T.M. and Walton, E.F. (1989) Carbohydrate requirements of peach fruit growth and respiration. *Tree Physiology* 5, 329–335.
Grossman, Y.L. and DeJong, T.M. (1994) Carbohydrate requirements for dark respiration by peach vegetative organs. *Tree Physiology* 14, 37–48.
Hansen, P. (1970) C-studies on apple trees. VI. The influence of the fruit on photosynthesis of the leaves, and the relative photosynthetic yields of fruits and leaves. *Physiologia Plantarum* 23, 805–810.
Lampinen, B.D., Shackel, K.A., Southwick, S.M., Olson, W.H. and DeJong, T.M. (2004) Leaf and canopy photosynthetic responses of French prune (*Prunus domestica* L. 'French') to stem water potential based deficit irrigation. *Journal of Horticultural Science and Biotechnology* 79, 638–644.

Walton, E.F. and DeJong, T.M. (1990) Estimating the bioenergetic cost of a developing kiwifruit berry and its growth and maintenance respiration components. *Annals of Botany* 66, 417–424.

Wünsche, J.N., Greer, D.H., Laing, W.A. and Palmer, J.W. (2005) Physiological and biochemical leaf and tree structure to crop load in apple. *Tree Physiology* 25, 1253–1263.

Figure sources

Taiz, L., Zeiger, E., Moller, I.M. and Murphy, A. (2015) *Plant Physiology and Development*. Sinauer Associates, Inc., Sunderland, Connecticut. (**Fig. 2.1**)

DeJong, T.M. (1989) Photosynthesis and respiration. In: LaRue, J.H. and Johnson, R.S. (eds) *Peaches, Plums, and Nectarines: Growing and Handling for Fresh Market*. Publication No. 3331, Division of Agriculture and Natural Resources, University of California, Oakland, California, pp. 38–41. (**Fig. 2.3**, **Fig. 2.7**)

DeJong, T.M., Shackel, K.A. and Lampinen, B.D. (2012) Carbohydrate assimilation and water use. In: Buchner, R.P. (ed.) *Prune Production Manual*. Publication No. 3507, Division of Agriculture and Natural Resources, University of California, Oakland, California, pp. 25–32. (**Fig. 2.4**)

Rosati, A., Esparza, G., DeJong, T.M. and Pearcy, R.W. (1999) Influence of canopy light environment and nitrogen availability on leaf photosynthetic characteristics and photosynthetic nitrogen-use efficiency of field-grown nectarine trees. *Tree Physiology* 19, 173–180. (**Fig. 2.5**)

Rosati, A., DeJong, T.M. and Esparza, G. (2002) Physiological basis for light use efficiency models. *Acta Horticulturae* 584, 89–94. (**Fig. 2.6**)

Lampinen, B.D., Shackel, K.A., Southwick, S.M., Olson, W.H. and DeJong, T.M. (2004) Leaf and canopy photosynthetic responses of French prune (*Prunus domestica* L. 'French') to stem water potential based deficit irrigation. *Journal of Horticultural Science and Biotechnology* 79, 638–644. (**Fig. 2.8**, **Fig. 2.9**)

Auzmendi, I., Marsal, J., Girona, J. and Lopez, G. (2013) Daily photosynthetic radiation use efficiency for apple and pear leaves: seasonal changes and estimation of canopy net carbon exchange rate. *European Journal of Agronomy* 51, 1–8. (**Fig. 2.10**)

Grossman, Y.L. and DeJong, T.M. (1994) Carbohydrate requirements for dark respiration by peach vegetative organs. *Tree Physiology* 14, 37–48. (**Fig. 2.11**)

Chapter 3: Uptake and Assimilation of Nutrient Resources

Most of the concepts related to uptake and assimilation of nutrient resources presented in this chapter are commonly described in modern plant physiology textbooks such as:

Taiz, L., Zaiger, E., Moller, I.M. and Murphy, A. (2015) *Plant Physiology and Development*. Sinauer Associates, Inc., Sunderland, Connecticut.

More detailed information on this topic related to fruit trees can be found in:

Brown, P.H. and Hu, H. (1996) Phloem mobility of boron is species dependent. Evidence for phloem mobility in sorbitol rich species. *Annals of Botany* 77, 497–505.

Castagnoli, S.P., Weinbaum, S.A. and DeJong, T.M. (1989) Nitrogen remobilization and nursery tree growth following autumn defoliation in plum (*Prunus salicina*) trees. *Advances in Horticultural Science* 3, 115–119.

Castagnoli, S.P., DeJong, T.M., Weinbaum, S.A. and Johnson, R.S. (1990) Autumn foliage applications of $ZnSO_4$ reduced leaf nitrogen remobilization in peach and nectarine. *Journal of the American Society for Horticultural Science* 115(1), 79–83.

DeJong, T.M. and Doyle, J.F. (1985) Seasonal relationships between leaf nitrogen content (photosynthetic capacity) and leaf canopy light exposure in peach (*Prunus persica*). *Plant, Cell & Environment* 8, 701–706.

DeJong, T.M., Day, K.R. and Johnson, R.S. (1989) Partitioning of leaf nitrogen with respect to within canopy light exposure and nitrogen availability in peach (*Prunus persica*). *Trees* 3, 89–95.

Deng, X., Weinbaum, S.A. and DeJong, T.M. (1989a) Use of labeled nitrogen to monitor transition in nitrogen dependence from storage to current-year uptake in mature walnut trees. *Trees* 3, 11–16.

Deng, X., Weinbaum, S.A., DeJong, T.M. and Muraoka, T.T. (1989b) Utilization of nitrogen from storage and current-year uptake in walnut spurs during the spring flush of growth. *Physiologia Plantarum* 75, 492–498.

Epstein, E. and Bloom, A.J. (2004) *Mineral Nutrition of Plants: Principles and Perspectives*. Sinauer Associates, Inc., Sunderland, Connecticut.

Johnson, R.S. and Day, K.R. (2013) Zinc deficiency and correction in California plum orchards. *Acta Horticulturae* 985, 189–191.

Muhammad, S., Sanden, B.L., Lampinen, B., Saa, S., Muhammad, I., et al. (2015) Seasonal changes in nutrient content and concentrations in a mature deciduous tree species: studies in almond (*Prunus dulcis* (Mill.) D. A. Webb). *European Journal of Agronomy* 65, 52–68. https://doi.org/10.1016/j.eja.2015.01.004

Muhammad, S., Sanden, B.L., Lampinen, B.D., Smart, D.R., Saa, S., et al. (2020) Nutrient storage in the perennial organs of deciduous trees and remobilization in spring – a study in almond (*Prunus dulcis*) (Mill.) D. A. Webb. *Frontiers in Plant Science* 11, 658. https://doi.org/10.3389/fpls.2020.00658

Rosati, A., Day, K.R. and DeJong, T.M. (2000) Distribution of leaf mass per unit area and leaf nitrogen concentration determine partitioning of leaf nitrogen within tree canopies. *Tree Physiology* 20, 271–276.

Rufat, J. and DeJong, T.M. (2001) Estimating seasonal nitrogen dynamics in peach trees in response to nitrogen availability. *Tree Physiology* 21, 1133–1140.

Ziska, L.H., DeJong, T.M., Hoffman, G.F. and Mead, R.M. (1991) Sodium and chloride distribution in salt-stressed *Prunus salicina*, a deciduous tree species. *Tree Physiology* 8, 47–57.

Figure and table sources

Taiz, L., Zeiger, E., Moller, I.M. and Murphy, A. (2015) *Plant Physiology and Development*. Sinauer Associates, Inc., Sunderland, Connecticut. (**Fig. 3.1**)

Muhammad, S., Sanden, B.L., Lampinen, B.D., Smart, D.R., Saa, S., *et al.* (2020) Nutrient storage in the perennial organs of deciduous trees and remobilization in spring – a study in almond (*Prunus dulcis*) (Mill.) D. A. Webb. *Frontiers in Plant Science* 11, 658. https://doi.org/10.3389/fpls.2020.00658 (**Table 3.1**)

Chapter 4: The Structure of Trees

Most of the concepts discussed in this chapter can be found in:

Kramer, P.J. and Kozlowski, T.T. (1979) *Physiology of Woody Plants*. Academic Press, New York.
Wilson, B.F. (1970) *The Growing Tree*. University of Massachusetts Press, Amherst, Massachusetts.
Zimmerman, M.H. and Brown, C.L. (1971) *Trees: Structure and Function*. Springer-Verlag, New York.

Figure sources

Rost, T.L., Barbour, M.G., Stocking, R. and Murphy, T.M. (2014) Plant Biology (Free Access Edition. Copyright Terence M. Murphy, Thomas L. Rost, Michael G. Barbour 2014. All federal and state copyrights reserved). Available at: https://labs.plb.ucdavis.edu/courses/bis/1c/text/PLANTBIOLOGY1.htm (accessed 31 August 2021). (**Fig. 4.1**, **Fig. 4.2**)

Chapter 5: The Carbohydrate Economy of Fruit Trees

The concept of plants being viewed as a collection of semi-autonomous organs was adopted from the following papers:

Watson, M.A. and Casper, B.B. (1984) Morphogenetic constraints on patterns of carbon distribution in plants. *Annual Review of Ecology and Systematics* 15, 233–258.
White, J. (1979) The plant as a metapopulation. *Annual Review of Ecology and Systematics* 10, 109–145.

This concept was used as a basis for modeling the distribution of carbon in fruit trees by Grossman and DeJong (1994) and later formalized by DeJong (1999):

Grossman, Y.L. and DeJong, T.M. (1994) PEACH: a simulation model of reproductive and vegetative growth in peach trees. *Tree Physiology* 14, 329–345.
DeJong, T.M. (1999) Developmental and environmental control of dry-matter partitioning in peach. *HortScience* 34(6), 1037–1040.

Additional studies that influenced the concepts presented on this subject are the following:

Berman, M.E. and DeJong, T.M. (2003) Seasonal patterns of vegetative growth and competition with reproductive sinks in peach (*Prunus persica*). *Journal of Horticultural Science and Biotechnology* 78(3), 303–309.

DeJong, T.M. (2017) The understanding of carbohydrate budgets in fruit trees made easy: what we know and ideas about what we need to know. *Acta Horticulturae* 1177, 29–39.

DeJong, T.M. and Grossman, Y.L. (1995) Quantifying sink and source limitations on dry matter partitioning to fruit growth in peach trees. *Physiologia Plantarum* 95, 437–443.

Marsal, J., Basile, B., Solari, L. and DeJong, T.M. (2003) Influence of branch autonomy on fruit, scaffold, trunk and root growth during Stage III of peach fruit development. *Tree Physiology* 23, 313–323.

Figure sources

DeJong, T.M. (1989) Photosynthesis and respiration. In: LaRue, J.H. and Johnson, R.S. (eds) *Peaches, Plums, and Nectarines: Growing and Handling for Fresh Market.* Publication No. 3331, Division of Agriculture and Natural Resources, University of California, Oakland, California, pp. 38–41. (**Fig. 5.1**)

DeJong, T.M., Doyle, J.F. and Day, K.R. (1987) Seasonal patterns of reproductive and vegetative sink activity in early and late maturing peach (*Prunus persica*) cultivars. *Physiologia Plantarum* 71, 83–88. (**Fig. 5.3**)

Marsal, J., Basile, B., Solari, L. and DeJong, T.M. (2003) Influence of branch autonomy on fruit, scaffold, trunk and root growth during Stage III of peach fruit development. *Tree Physiology* 23, 313–323. (**Fig. 5.4**)

Chapter 6: Understanding the Shoot Sink

A major activity in the management of fruit trees is pruning. To better understand responses to pruning it was necessary to conduct studies related to shoot growth and responses to pruning. Because of my views on carbon distribution it was important to understand competition between shoot growth and fruit growth and how shoot growth is regulated. Many of the concepts about shoot growth were from *The Growing Tree* by B.F. Wilson (1970) and the general refences cited for Chapter 4.

Additional concepts came from experience in studies of fruit tree shoot growth reported in the following publications:

Berman, M.E. and DeJong, T.M. (1997a) Diurnal patterns of stem extension growth in peach (*Prunus persica*): temperature and fluctuations in water status determine growth rate. *Physiologia Plantarum* 100, 361–370.

Berman, M.E. and DeJong, T.M. (1997b) Crop load and water stress effects on daily stem growth in peach (*Prunus persica*). *Tree Physiology* 17, 467–472.

Costes, E., Lauri, P.E. and Regnard, J.L. (2006) Analyzing fruit tree architecture: implications for tree management and fruit production. In: Janick, J. (ed.) *Horticultural Reviews*, Vol. 32. Wiley, Hoboken, New Jersey, pp. 1–61.

Davidson, A., Da Silva, D., Quintana, B. and DeJong, T.M. (2015) The phyllochron of *Prunus persica* shoots is relatively constant under controlled growth conditions but seasonally increases in the field in ways unrelated to patterns of temperature or radiation. *Scientia Horticulturae* 184, 106–113.

Davidson A., Da Silva, D. and DeJong, T.M. (2017) The phyllochron of well-watered and water deficit mature peach trees varies with shoot type and vigour. *AoB PLANTS* 9, plx042. https://doi.org/10.1093/aobpla/plx042

Davidson, A., Da Silva, D. and DeJong, T.M. (2019) Rate of shoot development (phyllochron) is dependent on carbon availability, shoot type, and rank in peach trees. *Trees* 33, 1583–1590. https://doi.org/10.1007/s00468-019-01881-y

DeJong, T.M., Doyle, J.F. and Day, K.R. (1987) Seasonal patterns of reproductive and vegetative sink activity in early and late maturing peach (*Prunus persica*) cultivars. *Physiologia Plantarum* 71, 83–88.

Fyhrie, K., Prats-Llinàs, M.T., López, G. and DeJong, T.M. (2018) How does peach fruit set on sylleptic shoots borne on epicormics compare with fruit set on proleptic shoots? *European Journal of Horticultural Science* 03, 3–11.

Gordon, D. and DeJong, T.M. (2007) Current-year and subsequent-year effects of crop load manipulation and epicormic-shoot removal on distribution of long, short and epicormic shoot growth in *Prunus persica*. *Annals of Botany* 99, 323–332.

Grossman, Y.L. and DeJong, T.M. (1995) Maximum vegetative growth potential and seasonal patterns of resource dynamics during peach growth. *Annals of Botany* 76, 473–482.

Negron, C., Contador, L., Lampinen, B.D., Metcalf, S.G., Guedon, Y., *et al.* (2013) Systematic analysis of branching patterns of three almond cultivars with different tree architectures. *Journal of the American Society for Horticultural Science* 138, 407–415.

Negron, C., Contador, L., Lampinen, B.D., Metcalf, S.G., Guedon, Y., *et al.* (2014) Differences in proleptic and epicormic shoot structures in relation to water deficit and growth rate in almond trees (*Prunus dulcis*). *Annals of Botany* 113, 545–554.

Prats-Llinàs, M.T., López, G., Fyrie, K., Pallas, B., Guédon, Y., *et al.* (2019) Long proleptic and sylleptic shoots in peach (*Prunus persica* L. Batsch) trees have similar, predetermined, maximum numbers of nodes and bud fate patterns. *Annals of Botany* 123, 993–1004.

Spann, T.M., Beede, R.H. and DeJong, T.M. (2007) Preformation in vegetative buds of pistachio (*Pistacia vera*): relationship to shoot morphology, crown structure and rootstock vigor. *Tree Physiology* 27, 1189–1196.

Sperling, O., Kamai, T., Tixier, A., Davidson, A., Jarvis-Shean, K., *et al.* (2019) Predicting bloom dates by temperature mediated kinetics of carbohydrate metabolism in deciduous fruit trees. *Agricultural and Forest Meteorology* 276–277, 107643. https://doi.org/10.1016/j.agrformet.2019.107643

Figure sources

Negron, C., Contador, L., Lampinen, B.D., Metcalf, S.G., Guedon, Y., *et al.* (2014) Differences in proleptic and epicormic shoot structures in relation to water deficit and growth rate in almond trees (*Prunus dulcis*). *Annals of Botany* 113, 545–554. (**Fig. 6.5**)

Wilson, B.F. and Howard, R.A. (1968) A computer model for vascular activity. *Forest Science* 14, 77–90. (**Fig. 6.8**)

Costes, E., Lauri, P.E. and Regnard, J.L. (2006) Analyzing fruit tree architecture: implications for tree management and fruit production. In: Janick, J. (ed.) *Horticultural Reviews*, Vol. 32. Wiley, Hoboken, New Jersey, pp. 1–61. (**Fig. 6.9**)

Negron, C., Contador, L., Lampinen, B.D., Metcalf, S.G., Guedon, Y., *et al.* (2013) Systematic analysis of branching patterns of three almond cultivars with different tree architectures. *Journal of the American Society for Horticultural Science* 138, 407–415. (**Fig. 6.11**)

Marsal, J., Basile, B., Solari, L. and DeJong, T.M. (2003) Influence of branch autonomy on fruit, scaffold, trunk and root growth during Stage III of peach fruit development. *Tree Physiology* 23, 313–323. (**Fig. 6.12**)

Chapter 7: Application of Shoot Growth Rules for Understanding Responses to Pruning

Studying the behavior of shoot growth and the desire to apply concepts regarding shoot growth resulted in numerous attempts to develop practical applications of the concepts described in Chapter 6 for understanding and managing fruit trees in orchard systems.

Examples of these studies are listed below:

DeJong, T.M. and Day, K.R. (1991) Relationships between shoot productivity and leaf characteristics in peach canopies. *HortScience* 26(10), 1271–1273.

DeJong, T.M., Day, K.R. and Doyle, J.F. (1992) Evaluation of training/pruning systems for peach, plum and nectarine trees in California. *Acta Horticulturae* 322, 99–105.

DeJong, T.M., Day, K.R., Doyle, J.F. and Johnson, R.S. (1994) The Kearney Agricultural Center Perpendicular "V" (KAC-V) orchard system for peaches and nectarine. *HortTechnology* 4(4), 362–367.

DeJong, T.M., Tsuji, W., Doyle, J.F. and Grossman, Y.L. (1999) Comparative economic efficiency of four peach product systems in California. *HortScience* 34(1), 73–78.

DeJong, T.M., Negron, C., Favreau, R., Costes, E., Lopez, G., *et al.* (2012) Using concepts of shoot growth and architecture to understand and predict responses of peach trees to pruning. *Acta Horticulturae* 962, 225–232.

Grossman, Y.L. and DeJong, T.M. (1998) Training and pruning system effects on vegetative growth potential, light interception, and cropping efficiency in peach trees. *Journal of the American Society for Horticultural Science* 123(6), 1058–1064.

Negron, C., Contador, L., Lampinen, B.D., Metcalf, S.G., Guedon, Y., *et al.* (2015) How different pruning severities alter shoot structure: a modelling approach in young 'Nonpareil' almond trees. *Functional Plant Biology* 42, 325–335.

Spann, T.M., Beede, R.H. and DeJong, T.M. (2008) Neoformed growth responses to dormant pruning in mature and immature pistachio trees grown on different rootstocks. *Journal of Horticultural Science and Biotechnology* 83, 137–142.

Chapter 8: Understanding the Root Sink

Many of the general concepts about roots described in this chapter came from a chapter on roots in the following book:

Rom, R.C. and Carlson, R.F. (eds) (1987) *Rootstocks for Fruit Crops*. Wiley, New York.

Most of the concepts and theories related to the influence of size-controlling rootstocks on tree growth and physiology were influenced by the following studies conducted in association with my research group:

Basile, B. and DeJong, T.M. (2019) Control of fruit tree vigor induced by dwarfing rootstocks. *Horticultural Reviews* 46, 39–98.

Basile, B., Marsal, J. and DeJong, T.M. (2003a) Daily shoot extension growth of peach trees growing on rootstocks that reduce scion growth is related to daily dynamics of stem water potential. *Tree Physiology* 23, 695–704.

Basile, B., Marsal, J., Solari, L.I., Tyree, M.T., Bryla, D.R. and DeJong, T.M. (2003b) Hydraulic conductance of peach trees grafted on rootstocks with differing size-controlling potentials. *Journal of Horticultural Science and Biotechnology* 78(5), 768–774.

Basile, B., Bryla, D.R., Salsman, M.L., Marsal, J., Cirillo, C., *et al.* (2007) Growth patterns and morphology of fine roots of size-controlling and invigorating peach rootstocks. *Tree Physiology* 27, 231–241.

Ben Mimoun, M. and DeJong, T.M. (2006) Effect of fruit crop load on peach root growth. *Acta Horticulturae* 713, 169–175.

Bruckner, C.H. and DeJong, T.M. (2014) Proposed pre-selection method for identification of dwarfing peach rootstocks based on rapid shoot xylem vessel analysis. *Scientia Horticulturae* 165, 404–409.

DeJong, T.M., Tombesi, S., Basile, B. and Da Silva, D. (2013) Beakbane and Thompson (1939, East Malling) had it right: scion vigour is physiologically linked to the xylem anatomy of the rootstock. *Aspects of Applied Biology* 119, 51–58.

Solari, L.I. and DeJong, T.M. (2006) The effect of root pressurization on water relations, shoot growth, and leaf gas exchanges of peach (*Prunus persica*) trees on rootstocks with differing growth potential and hydraulic conductance. *Journal of Experimental Botany* 57, 1981–1989.

Solari, L.I., Johnson, S. and DeJong, T.M. (2006a) Relationship of water status to vegetative growth and leaf gas exchange of peach (*Prunus persica*) trees on different rootstocks. *Tree Physiology* 26, 1333–1341.

Solari, L.I., Johnson, S. and DeJong, T.M. (2006b) Hydraulic conductance characteristics of peach (*Prunus persica*) trees on different rootstocks are related to biomass production and distribution. *Tree Physiology* 26, 1343–1350.

Solari, L.I., Pernice, F. and DeJong, T.M. (2006c) The relationship of hydraulic conductance to root system characteristics of peach (*Prunus persica*) rootstocks. *Physiologia Plantarum* 128, 324–333.

Tombesi S., Johnson, R.S., Day, K.R. and DeJong, T.M (2010a) Interactions between rootstock, inter-stem and scion xylem vessel characteristics of peach trees growing on rootstocks with contrasting size-controlling characteristics. *AoB PLANTS* 2010, plq013.

Tombesi, S., Johnson, R.S., Day, K.R. and DeJong, T.M. (2010b) Relationships between xylem vessel characteristics, calculated axial hydraulic conductance and size-controlling capacity of peach rootstocks. *Annals of Botany* 105, 327–331.

Tombesi, S., Almehdi, A. and DeJong, T.M. (2010c) Phenotyping vigour control capacity of new peach rootstocks by xylem vessel analysis. *Scientia Horticulturae* 127, 353–357.

Weibel, A., Johnson, R.S. and DeJong, T.M. (2003) Comparative vegetative growth responses of two peach cultivars grown on size-controlling versus standard rootstocks. *Journal of the American Society for Horticultural Science* 128(4), 463–471.

Figure sources

Taiz, L., Zaiger, E., Moller, I.M. and Murphy, A. (2015) *Plant Physiology and Development*. Sinauer Associates, Inc., Sunderland, Connecticut. (**Fig. 8.1**)

Ben Mimoun, M. and DeJong, T.M. (2006) Effect of fruit crop load on peach root growth. *Acta Horticulturae* 713, 169–175. (**Fig. 8.3**)

Marsal, J., Basile, B., Solari, L. and DeJong, T.M. (2003) Influence of branch autonomy on fruit, scaffold, trunk and root growth during Stage III of peach fruit development. *Tree Physiology* 23, 313–323. (**Fig. 8.4**)

Basile, B. and DeJong, T.M. (2019) Control of fruit tree vigor induced by dwarfing rootstocks. *Horticultural Reviews* 46, 39–98. (**Fig. 8.5**)

Chapter 9: Understanding the Fruit Sink

The double-sigmoid curve of many stone fruit species has been the subject of many pomological studies for more than 100 years. Many theories have been proposed to explain the cause of this pattern, but in 1986 I initiated a study that resulted in showing that the double-sigmoid patterns of accumulative fruit weight increases were simply a mathematical result of a rapidly descending fruit relative growth rate function:

DeJong, T.M. and Goudriaan, J. (1989) Modeling peach fruit growth and carbohydrate requirements: reevaluation of the double-sigmoid growth pattern. *Journal of the American Society for Horticultural Science* 114(5), 800–804.

Subsequent studies investigated numerous aspects of fruiting and fruit development that have influenced the contents of this chapter on the fruit as a sink for assimilates. These include:

Basile, B., Mariscal, M.J., Day, K.R., Johnson, R.S. and DeJong, T.M. (2002) Japanese plum (*Prunus salicina* L.) fruit growth: seasonal pattern of source/sink limitations. *Journal of the American Pomological Society* 56(2), 86–93.

Berman, M.E. and DeJong, T.M. (1996) Water stress and crop load effects on fruit fresh and dry weights in peach (*Prunus persica*). *Tree Physiology* 16, 859–864.

Berman, M.E., Rosati, A., Pace, L., Grossman, Y.L. and DeJong, T.M. (1998) Using simulation modeling to estimate the relationship between date of fruit maturity and yield potential in peach. *Fruit Varieties Journal* 52(4), 229–235.

DeJong, T.M. (2006) Physiological and developmental principles of peach tree and fruit growth related to management practices. *Acta Horticulturae* 713, 161–168.

DeJong, T.M. (2012) Fruit growth and development as it relates to crop load, thinning and climate change. *Acta Horticulturae* 962, 233–238.

DeJong, T.M. and Walton, E.F. (1989) Carbohydrate requirements of peach fruit growth and respiration. *Tree Physiology* 5, 329–335.

Fyhrie, K., Prats-Llinàs, M.T., López, G. and DeJong, T.M. (2018) How does peach fruit set on sylleptic shoots borne on epicormics compare with fruit set on proleptic shoots? *European Journal of Horticultural Science* 83, 3–11.

Grossman, Y.L. and DeJong, T.M. (1995a) Maximum fruit growth potential and seasonal patterns of resource dynamics during peach growth. *Annals of Botany* 75, 553–560.

Grossman, Y.L. and DeJong, T.M. (1995b) Maximum fruit growth potential following resource limitation during peach growth. *Annals of Botany* 75, 561–567.

Klein, I., Weinbaum, S.A., DeJong, T.M. and Muraoka, T.T. (1991) Relationship between fruiting, specific leaf weight, and subsequent spur productivity in walnut. *Journal of the American Society for Horticultural Science* 116(3), 426–429.

Lopez, G. and DeJong, T.M. (2007) Spring temperatures have a major effect on early stages of peach fruit growth. *Journal of Horticultural Science and Biotechnology* 82, 507–512.

Pavel, E.W. and DeJong, T.M. (1993a) Seasonal CO_2 exchange patterns of developing peach (*Prunus persica*) fruits in response to temperature, light and CO_2 concentration. *Physiologia Plantarum* 88, 322–330.

Pavel, E.W. and DeJong, T.M. (1993b) Estimating the photosynthetic contribution of developing peach (*Prunus persica*) fruits to their growth and maintenance carbohydrate requirements. *Physiologia Plantarum* 88, 331–338.

Pavel, E.W. and DeJong, T.M. (1993c) Relative growth rate and its relationship to compositional changes of nonstructural carbohydrates in the mesocarp of developing peach fruits. *Journal of the American Society for Horticultural Science* 118(4), 503–508.

Pavel, E.W. and DeJong, T.M. (1993d) Source- and sink-limited growth periods of developing peach fruits indicated by relative growth rate analysis. *Journal of the American Society for Horticultural Science* 118(6), 820–824.

Pavel, E.W. and DeJong, T.M. (1995) Seasonal patterns of nonstructural carbohydrates of apple (*Malus pumila* Mill.) fruits: relationship with relative growth rates and contribution to solute potential. *Journal of Horticultural Science* 70(1), 127–134.

Saenz, J.L., DeJong, T.M. and Weinbaum, S.A. (1997) Nitrogen stimulated increases in peach yields are associated with extended fruit development period and increased fruit sink capacity. *Journal of the American Society for Horticultural Science* 122(6), 772–777.

Tombesi, S., Scalia, R., Connell, J., Lampinen, B. and DeJong, T.M. (2010) Fruit development in almond is influenced by early spring temperatures in California. *Journal of Horticultural Science and Biotechnology* 85, 317–322.

Walton, E.F. and DeJong, T.M. (1990a) Estimating the bioenergetic cost of a developing kiwifruit berry and its growth and maintenance respiration components. *Annals of Botany* 66, 417–424.

Walton, E.F. and DeJong, T.M. (1990b) Growth and compositional changes in kiwifruit berries from three Californian locations. *Annals of Botany* 66, 285–298.

Walton, E.F., DeJong, T.M. and Loomis, R.S. (1990) Comparison of four methods calculating the seasonal pattern of plant growth efficiency of a kiwifruit berry. *Annals of Botany* 66, 299–307.

Figure and table sources

Lamp, B.M., Connell, J.H., Duncan, R.A., Viveros, M. and Polito, V.S. (2001) Almond flower development: floral initiation and organogenesis. *Journal of the American Society for Horticultural Science* 126, 689–696. (**Fig. 9.1**, **Fig. 9.2**)

Polito, V.S., DeCeault, M. and Norton, M.V. (2012) Setting the crop: flowering, pollination and fruit development. In: Buchner, R.P. (ed.) *Prune Production Manual*. Publication No. 3507, Division of Agriculture and Natural Resources, University of California, Oakland, California, pp. 17–32. (**Fig. 9.4a**)

Polito, V.S. (1998) Floral biology: flower structure, development and pollination. In: Ramos, D.E. (ed.) *Walnut Production Manual*. Publication No. 3373, Division of Agriculture and Natural Resources, University of California, Oakland, California, pp. 127–132. (**Fig. 9.4b**)

Westwood, M.N. (1978) *Temperate Zone Pomology*. W.H. Freeman and Company, San Francisco, California. (**Fig. 9.6**, **Fig. 9.8**, **Fig. 9.9**)

UC Davis Fruit and Nut Research and Information Center (2021) Fruit anatomy. Available at: http://fruitandnuteducation.ucdavis.edu/generaltopics/Anatomy Pollination/Fruit_Anatomy/ (accessed 31 August 2021). (**Fig. 9.7**)

Grossman, Y.L. and DeJong, T.M. (1995a) Maximum fruit growth potential and seasonal patterns of resource dynamics during peach growth. *Annals of Botany* 75, 553–560 (**Fig. 9.10**, **Fig. 9.11**, **Fig. 9.12**)

Pavel, E.W. and DeJong, T.M. (1995) Seasonal patterns of nonstructural carbohydrates of apple (*Malus pumila* Mill.) fruits: relationship with relative growth rates and contribution to solute potential. *Journal of Horticultural Science* 70(1), 127–134. (**Fig. 9.12**)

Lopez, G., Favreau, R.R., Smith, C. and DeJong, T.M. (2010) L-PEACH: a computer-based model to understand how peach trees grow. *HortTechnology* 20, 983–990 (**Fig. 9.13**)

Grossman, Y.L. and DeJong, T.M. (1995b) Maximum fruit growth potential following resource limitation during peach growth. *Annals of Botany* 75, 561–567. (**Fig. 9.14**)

Ben Mimoun, M. and DeJong, T.M. (1999) Using the relation between growing degree hours and harvest date to estimate run-times for peach: a tree growth and yield simulation model. *Acta Horticulturae* 499, 107–114. (**Fig. 9.15**)

Hawker, J.S. and Buttrose, M.S. (1980) Development of the almond nut (*Prunus dulcis* (Mill.) D. A. Webb). Anatomy and chemical composition of fruit parts from anthesis to maturity. *Annals of Botany* 46, 313–321. (**Fig. 9.16**)

Pinney, K., Labavitch, J.M. and Polito, V.S. (1998) Fruit growth and development. In: Ramos, D.E. (ed.) *Walnut Production Manual*. Publication No. 3373, Division of Agriculture and Natural Resources, University of California, Oakland, California, pp. 139–143. (**Fig. 9.18**)

Geisseler, D. and Horwath, W.R. (2016) Pistachio Production in California. California Department of Food and Agriculture Fertilizer Research and Education Program (FREP) public access PDF. Available at: https://apps1.cdfa.ca.gov/FertilizerResearch/docs/Pistachio_Production_CA.pdf (accessed 26 August 2021). (**Fig. 9.21**)

Tombesi, S., Lampinen, B.D., Metcalf, S. and DeJong, T.M. (2016) Yield in almond is related more to the abundance of flowers than the relative number of flowers that set fruit. *California Agriculture* 71, 68–74. (**Fig. 9.22**)

DeJong, T.M. (2012) Fruit growth and development as it relates to crop load, thinning and climate change. *Acta Horticulturae* 962, 233–238. (**Table 9.1**)

Chapter 10: Understanding the Long-Term Storage Sink

The general pattern of long-term carbohydrate storage presented in this chapter was influenced by reports in general references such as:

Kozlowski, T.T., Kramer, P.J. and Pallardy, S.G. (1991) *The Physiological Ecology of Woody Plants*. Academic Press, San Diego, California.

Priestley, C.A. (1970) Carbohydrate storage and utilization. In: Luckwill, L.C. and Cutting, C.V. (eds) *Physiology of Tree Crops*. Academic Press, London, pp. 113–127.

The concept of long-term storage being driven by an active carbohydrate sink came from:

Cannell, M.G.R. and Dewar, R.C. (1994) Carbon allocation in trees: a review of concepts for modeling. In: Begon, M. and Fitter, A.H. (eds) *Advances in Ecological Research*, Vol. 25. Academic Press, London, pp. 59–104.

This concept was applied to the mechanistic modeling of carbohydrate storage in fruit trees in:

Da Silva, D., Qin, L., DeBuse, C. and DeJong, T.M. (2014) Measuring and modelling seasonal patterns of carbohydrate storage and mobilization in the trunks and root crowns of peach trees. *Annals of Botany* 114, 643–652.

The following is additional research on this subject conducted in association with my group:

DeJong, T.M. (2016) Demystifying carbohydrate allocation to storage in fruit trees. *Acta Horticulturae* 1130, 329–334.

Kirschbaum D.S., Larson, K.D., Weinbaum, S.E. and DeJong, T.M. (2010) Relationships of carbohydrate and nitrogen content with strawberry transplant vigor and fruiting pattern in annual production systems. *The Americas Journal of Plant Science and Biotechnology* 4, 98–103.

Kirschbaum, D.S., Larson, K.D., Weinbaum, S.A. and DeJong, T.M. (2012) Accumulation pattern of total nonstructural carbohydrate in strawberry runner plants and its influence on plant growth and fruit production. *African Journal of Biotechnology* 11, 16253–16262.

Larson, K.D., DeJong, T.M. and Johnson, R.S. (1988) Physiological and growth responses of mature peach trees to postharvest water stress. *Journal of the American Society for Horticultural Science* 113(3), 296–300.

Spann, T.M., Beede, R.H. and DeJong, T.M. (2008) Seasonal carbohydrate storage and mobilization in bearing and non-bearing pistachio (*Pistacia vera*) trees. *Tree Physiology* 28, 207–213.

Sperling, O., Kamai, T., Tixier, A., Davidson, A., Jarvis-Shean, K., *et al.* (2019) Predicting bloom dates by temperature mediated kinetics of carbohydrate metabolism in deciduous fruit trees. *Agricultural and Forest Meteorology* 276–277, 107643. https://doi.org/10.1016/j.agrformet.2019.107643

Figure and table sources

Zwieniecki, M. and Davidson, A. (n.d.) The UC Davis Carbohydrate Observatory. Available at: https://psfaculty.plantsciences.ucdavis.edu/plantsciences_faculty/zwieniecki/CR/cr.html (accessed 25 August 2021). (**Fig. 10.1**)

Da Silva, D., Qin, L., DeBuse, C. and DeJong, T.M. (2014) Measuring and modelling seasonal patterns of carbohydrate storage and mobilization in the trunks and root crowns of peach trees. *Annals of Botany* 114, 643–652. (**Fig. 10.2**)

Murneek, A.E. (1942) *Quantitative Distribution of Nitrogen and Carbohydrates in Apple Trees.* Research Bulletin No. 348, University of Missouri, College of Agriculture, Agricultural Experiment Station, Columbia, Missouri. (**Table 10.1**)

Chapter 11: Integration of Tree Source and Sink Activities

The approach to modeling fruit tree growth and physiology described in this chapter was initially influenced by studies such as:

Loomis, R.S., Rabbinge, R. and Ng, E. (1979) Explanatory models in crop physiology. *Annual Review of Plant Physiology* 30, 339–367.

Penning de Vries, F.W.T. and van Laar, H.H. (1982) *Simulation of Plant Growth and Crop Production.* Centre for Agricultural Publishing and Documentation (Pudoc), Wageningen, the Netherlands.

The modeling approach for adapting the early models to do virtual modeling of tree growth and physiology was based on L-systems presented by:

Prusinkiewicz, P. and Lindenmayer, A. (1990) *The Algorithmic Beauty of Plants.* Springer-Verlag, New York.

The most important papers for understanding the 30-year progression of peach tree modeling are (in order of publication date):

DeJong, T.M. and Goudriaan, J. (1989) Modeling peach fruit growth and carbohydrate requirements: reevaluation of the double-sigmoid growth pattern. *Journal of the American Society for Horticultural Science* 114(5), 800–804.

Grossman, Y.L. and DeJong, T.M. (1994) PEACH: a simulation model of reproductive and vegetative growth in peach trees. *Tree Physiology* 14, 329–345.

Allen, M.T., Prusinkiewicz, P. and DeJong, T.M. (2005) Using L-systems for modeling source–sink interactions, architecture and physiology of growing trees: the L-PEACH model. *New Phytologist* 166, 869–888.

Allen, M.T., Prusinkiewicz, P., Favreau, R.R. and DeJong, T.M. (2007) L-PEACH, an L-system-based model for simulating architecture, carbohydrate source-sink interactions and physiological responses of growing trees. In: Vos, J., Marcelis, L., de Visser, P. and Struik, P. (eds) *Functional–Structural Plant Modelling in Crop Production*. Frontis, Wageningen, the Netherlands, pp. 139–150.

Prusinkiewicz, P., Allen, M., Escobar-Gutierrez, A. and DeJong, T.M. (2007) Numerical methods for transport–resistance sink–source allocation models. In: Vos, J., Marcelis, L., de Visser, P. and Struik, P. (eds) *Functional–Structural Plant Modelling in Crop Production*. Frontis, Wageningen, the Netherlands, pp. 123–138.

Lopez, G., Favreau, R.R., Smith, C., Costes, E., Prusinkiewicz, P. and DeJong, T.M. (2008) Integrating simulation of architectural development and source–sink behaviour of peach trees by incorporating Markov chains and physiological organ function submodels into L-PEACH. *Functional Plant Biology* 35, 761–771.

Da Silva, D., Favreau, R., Auzmendi, I. and DeJong, T.M. (2011) Linking water stress effects on carbon partitioning by introducing a xylem circuit into L-PEACH. *Annals of Botany* 108, 1135–1145.

Da Silva, D., Qin, L., DeBuse, C. and DeJong, T.M. (2014) Measuring and modelling seasonal patterns of carbohydrate storage and mobilization in the trunks and root crowns of peach trees. *Annals of Botany* 114, 643–652.

Additional papers on the subject of fruit crop modeling associated with the DeJong group are:

Allen, M.T., DeJong, T. and Prusinkiewicz, P. (2002) Using L-systems to model carbon transport and partitioning in developing peach trees. *Acta Horticulturae* 584, 29–34.

Allen, M., DeJong, T.M. and Prusinkiewicz, P. (2006) L-PEACH, an L-systems based model for simulating the architecture and carbon partitioning of growing fruit trees. *Acta Horticulturae* 707, 71–76.

Auzmendi, I., Hanan, J., Da Silva, D., Favreau, R. and DeJong, T.M. (2017) Modeling final leaf length as a function of carbon availability during the elongation period. *Acta Horticulturae* 1160, 75–81.

Ben Mimoun, M. and DeJong, T.M. (1999) Using the relation between growing degree hours and harvest date to estimate run-times for peach: a tree growth and yield simulation model. *Acta Horticulturae* 499, 107–114.

Berman, M.E., Rosati, A., Pace, L., Grossman, Y.L. and DeJong, T.M. (1998) Using simulation modeling to estimate the relationship between date of fruit maturity and yield potential in peach. *Fruit Varieties Journal* 52(4), 229–235.

Da Silva, D., Favreau, R.O., Tombesi, S. and DeJong, T.M. (2015) Modeling size-controlling rootstock effects on peach tree growth and development using L-PEACH-h. *Acta Horticulturae* 1068, 227–233.

DeJong, T.M. (2019) Opportunities and challenges in fruit tree and orchard modelling. *European Journal of Horticultural Science* 84, 117–123.

DeJong, T.M. and Goudriaan, J. (1989) Modeling the carbohydrate economy of peach fruit growth and crop production. *Acta Horticulturae* 254, 103–108.

DeJong, T.M. and Grossman, Y.L. (1992) Modelling the seasonal carbon economy of deciduous tree crops. *Acta Horticulturae* 313, 21–28.

DeJong, T.M. and Grossman, Y.L. (1994) A supply and demand approach to modeling annual reproductive and vegetative growth of deciduous fruit trees. *HortScience* 29(12), 1435–1442.

DeJong, T.M., Johnson, R.S. and Castagnoli, S.P. (1990) Computer simulation of the carbohydrate economy of peach crop growth. *Acta Horticulturae* 276, 97–104.

DeJong, T.M., Grossman, Y.L., Vosburg, S.F. and Pace, L.S. (1996) PEACH: a user friendly peach tree growth and yield simulation model for research and education. *Acta Horticulturae* 416, 199–206.

DeJong, T.M., Favreau, R., Allen, M. and Prusinkiewicz, P. (2008) Using computer technology to study, understand and teach how trees grow. *Acta Horticulturae* 772, 143–150

DeJong, T.M., Da Silva, D., Negron, C., Cieslak, M. and Prusinkiewicz, P. (2017) The L-ALMOND model: a functional-structural virtual tree model of almond tree architectural growth, carbohydrate dynamics over multiple years. *Acta Horticulturae* 1160, 43–49.

Esparza, G., DeJong, T.M. and Grossman, Y.L. (1998) Modeling the vegetative and reproductive growth of almonds. *Acta Horticulturae* 470, 324–331.

Esparza, G., DeJong, T.M. and Grossman, Y.L. (1999) Modifying 'peach' to model the vegetative and reproductive growth of almonds. *Acta Horticulturae* 499, 91–98.

Grossman, Y.L. and DeJong, T.M. (1994) Carbohydrate requirements for dark respiration by peach vegetative organs. *Tree Physiology* 14, 37–48.

Lopez, G., Smith, C., Favreau, R. and DeJong, T. (2008) Using L-PEACH for dynamic simulation of source-sink behavior of peach trees: effects of date of thinning on fruit growth. *Acta Horticulturae* 803, 209–216.

Lopez, G., Favreau, R.R., Smith, C. and DeJong, T.M. (2010) L-PEACH: a computer-based model to understand how peach trees grow. *HortTechnology* 20, 983–990.

Lopez, G., Girona, J., Marsal, J. and DeJong, T.M. (2015) Developing a physiological basis for modeling peach canopy photosynthesis under water stress conditions. *Acta Horticulturae* 1068, 203–209.

Lopez, G., Da Silva, D., Auzmendi, I., Favreau, R.R. and DeJong, T.M. (2018a) Demonstrative simulations of L-PEACH: a computer based model to understand how peach trees grow. *Acta Horticulturae* 1228, 13–19.

Lopez, G., Negron, C., Cieslak, M., Costes, E., Da Silva, D. and DeJong, T.M. (2018b) Simulation of tree growth for three almond cultivars with contrasting architecture with the L-ALMOND model. *Acta Horticulturae* 1228, 29–35.

Mariscal, M.J., Day, K.R., Basile, B. and DeJong, T.M. (2002) Modeling the vegetative and reproductive growth of plums. *Acta Horticulturae* 584, 35–42.

Rufat, J. and DeJong, T.M. (1999) Modeled seasonal pattern of nitrogen requirements of mature, cropping peach trees (*Prunus persica* (L.) Batsch). *Acta Horticulturae* 499, 129–135.

Rufat, J. and DeJong, T.M. (2001) Estimating seasonal nitrogen dynamics in peach trees in response to nitrogen availability. *Tree Physiology* 21, 1133–1140.

Smith, C., Costes, E., Favreau, R., Lopez, G. and DeJong, T. (2008) Improving the architecture of simulated trees in L-PEACH by integrating Markov chains and responses to pruning. *Acta Horticulturae* 803, 201–208.

Walton, E.F. and DeJong, T.M. (1990) Estimating the bioenergetic cost of a developing kiwifruit berry and its growth and maintenance respiration components. *Annals of Botany* 66, 417–424.

Walton, E.F., DeJong, T.M. and Loomis, R.S. (1990) Comparison of four methods calculating the seasonal pattern of plant growth efficiency of a kiwifruit berry. *Annals of Botany* 66, 299–307.

Figure sources

DeJong, T.M. and Goudriaan, J. (1989) Modeling peach fruit growth and carbohydrate requirements: reevaluation of the double-sigmoid growth pattern. *Journal of the American Society for Horticultural Science* 114, 800–804. (**Fig. 11.1, Fig. 11.2**)

Grossman, Y.L. and DeJong, T.M. (1994) PEACH: a simulation model of reproductive and vegetative growth in peach trees. *Tree Physiology* 14, 329–345. (**Fig. 11.3, Fig. 11.4, Fig. 11.5, Fig. 11.6**)

Prats-Llinàs, M.T., López, G., Fyrie, K., Pallas, B., Guédon, Y., et al. (2019) Long proleptic and sylleptic shoots in peach (*Prunus persica* L. Batsch) trees have similar, predetermined, maximum numbers of nodes and bud fate patterns. *Annals of Botany* 123, 993–1004. (**Fig. 11.11**)

Da Silva, D., Favreau, R., Auzmendi, I. and DeJong, T.M. (2011) Linking water stress effects on carbon partitioning by introducing a xylem circuit into L-PEACH. *Annals of Botany* 108, 1135–1145. (**Fig. 11.13, Fig. 11.14**)

Da Silva, D., Favreau, R.O., Tombesi, S. and DeJong, T.M. (2015) Modeling size-controlling rootstock effects on peach tree growth and development using L-PEACH-h. *Acta Horticulturae* 1068, 227–233. (**Fig. 11.15**)

Da Silva, D., Qin, L., DeBuse, C. and DeJong, T.M. (2014) Measuring and modelling seasonal patterns of carbohydrate storage and mobilization in the trunks and root crowns of peach trees. *Annals of Botany* 114, 643–652. (**Fig. 11.16**)

Lopez, G., Negron, C., Cieslak, M., Costes, E., Da Silva, D. and DeJong, T.M. (2018) Simulation of tree growth for three almond cultivars with contrasting architecture with the L-ALMOND model. *Acta Horticulturae* 1228, 29–35. (**Fig. 11.17**)

Index

CABI – who we are and what we do

This book is published by **CABI**, an international not-for-profit organisation that improves people's lives worldwide by providing information and applying scientific expertise to solve problems in agriculture and the environment.

CABI is also a global publisher producing key scientific publications, including world renowned databases, as well as compendia, books, ebooks and full text electronic resources. We publish content in a wide range of subject areas including: agriculture and crop science / animal and veterinary sciences / ecology and conservation / environmental science / horticulture and plant sciences / human health, food science and nutrition / international development / leisure and tourism.

The profits from CABI's publishing activities enable us to work with farming communities around the world, supporting them as they battle with poor soil, invasive species and pests and diseases, to improve their livelihoods and help provide food for an ever growing population.

CABI is an international intergovernmental organisation, and we gratefully acknowledge the core financial support from our member countries (and lead agencies) including:

Discover more

To read more about CABI's work, please visit: **www.cabi.org**

Browse our books at: **www.cabi.org/bookshop**,
or explore our online products at: **www.cabi.org/publishing-products**

Interested in writing for CABI? Find our author guidelines here:
www.cabi.org/publishing-products/information-for-authors/

Printed and bound by CPI Group (UK) Ltd, Croydon, CR0 4YY

06/08/2024

14537079-0001